Roadtripping at the End of the World

Roadtripping at the End of the World

essays & interviews by
KOLLIBRI TERRE SONNENBLUME

MACSKA MOKSHA PRESS
Cliff, New Mexico, USA

Cover photos: Unpaved road to Toquima Caves in Nevada (front) and "End" sign in Fort Bragg, California (back) by the author. Author portrait by Jeanette Stolen.

Published in 2019 by Macska Moksha Press
PO Box 301
Cliff, New Mexico, 88028
macskamoksha.com

Library of Congress Cataloging-in-Publication Data

Sonnenblume, Kollibri terre, 1969-
Roadtripping at the End of the World / Kollibri terre Sonnenblume

ISBN: 978-0-9861881-8-3

1. Travel 2. Nature 3. Culture I. Sonnenblume, Kollibri terre, 1969-. II. Title

To truly know the world, look deeply within your own being;
to truly know yourself, take real interest in the world.
—Rudolf Steiner

Table of Contents

Introduction

This book is a collection of essays and interviews. The oldest piece is set in 2014, the year I quit farming and started living on the road. Most of my time was spent on the US American West Coast where I worked seasonally in northern California, foraged for wild foods in Oregon, wintered in the deserts of southern California, worked on book projects in the city of Portland and the county of Mendocino, and camped all over the place, taking thousands of photographs. In 2018, I took a cross-country roadtrip all the way to the East Coast and back. Though I had no permanent address during this period, and often slept in the back of my 1986 Toyota pickup truck, I never called myself "homeless"; I felt the term was best reserved for those in greater need than me. Regardless, this lifestyle placed me at the margins of society, a position that I did not mind and which granted me an outside perspective I valued.

2014 was also the hottest year on record up to that point. The four years that followed—2015, 2016, 2017 and 2018—all beat it. The current year—2019—looks set to do the same, and might even take first place. These circumstance have helped contribute to the sensation that "something's not right" which has been spreading in society at large. News headlines referring to Climate Change increasingly include phrases like "sooner than predicted." The theory that humans could be facing near-term extinction is gaining greater currency. Could "the end of the world" be near?

Maybe. That depends what you mean. When I was farming, I used to say: "Nothing is 'the end of the world' except the end of the world, and maybe even *that* isn't." By this, I meant that the figurative is not the literal and that the nature of the literal itself is mysterious. But certainly, business-as-usual can't go on much longer for purely logistical reasons;

we are consuming the "resources" of the planet faster than they can be renewed—things like top soil, fresh water, forests, fuel and rare earth minerals. The planet's greater ecology is suffering from pollution, habitat loss, and extinctions, all of which affect humanity. At some point—perhaps much sooner than most people suspect—we will be forced to change our lifestyles drastically, if we do not choose to do so voluntarily before that. However, at least for now in US American society, there is no impetus to make big changes. That means we are definitely setting ourselves up for a fall one way or another.

The period since 2014 has been increasingly marked by extremes and unpredictability not only in the weather but also in politics, economics, and culture. Witness the polarization, inequality and hatred that define life more and more. While people disagree about causes and responses, it is undeniable that we are witnessing serious social entropy.

I've been watching all of these trends and events with increasing scrutiny. I've also been learning about the act of watching itself—that is, about perception. Such personal lessons are not always easy to transfer but I will try to do so from time to time in the following pages.

The essays here cover a variety of topics including agriculture, history, botany, indigenous issues, wildtending, racism, political activism, US imperialism, renewable energy, gentrification, architecture, metaphysics and more. The theme that's never far out of sight is Climate Chaos.

All the interviews except one were in-person roadside conversations with people I met up with along the way. Their knowledge and experience are expressed in their own words here with only light editing. I remain grateful to each of them for sharing their time and ideas with me.

Kollibri terre Sonnenblume
Occupied Pomo territory, 8/8/2019

ACKNOWLEDGMENTS

Thank you for proofing and editing assistance to Patrick McCafferty, Joshua Jude, Nicole Patrice Hill and Deva. Any remaining errors are my own. I completed this book in the summer of 2019 when a farm injury put me out of work. During this spell, I was kept afloat with generous contributions of housing and financial support from friends both new and old including: the Beasley siblings, Bernie Rauch, Michael Kelly, Heather Campbell, Nancy Gaedke, Meg Caskey & Dan Man, Anne P. Hill and Travis Q. Penrod.

TECHNICAL DETAILS

This entire project was produced with open source software on a refurbished laptop from Free Geek (Portland, Oregon). Details:

- OS: Linux Mint 17.1 Cinnamon
- text editor: gedit 2.30.4
- word processing and layout: LibreOffice Writer 4.2.8.2
- cover: GIMP (GNU Image Manipulation Program) 2.8.22

Fonts—body: Forum; italics: *Arapey*; titles, headings, bold, cover: **Knorke**.

Your Money or Your Life: Can We Afford to Work for Peanuts?

In Orick, California, there is a giant wooden peanut—over twelve feet long, six feet tall and weighing nine tons—carved from a single chunk of old-growth redwood. It was sculpted by local loggers in 1978, during the Carter administration, and brought to Washington, D.C., to protest the proposed expansion of Redwood National Park. Their message: "It may be peanuts to you, but it's jobs to us." To the loggers' chagrin, the peanut was ignored and the Park expanded anyway, removing 48,000 acres from the reach of their saws. Now the rough-hewn sculpture sits unceremoniously at the south end of town, steadily wearing away under the effects of vandalism and the elements. No plaque tells its story; you have to know what you're looking for.

In the Spring of 2014, my farming partner, Clarabelle, and I unexpectedly lost our lease on a piece of land we were planning to farm. This was a harsh blow that simultaneously took away our livelihood and made us homeless. Not knowing what else to do with our unexpected free time, we hit the road to lick our wounds in a 1985 Toyota van (aka a "lunar lander"). Both of us are genuine lovers of plants, animals and nature, so the trip was a pilgrimage of sorts and we hoped it would bring us some joy. But the condition of the environment being what it is, we also experienced many sad revelations.

Our travels took us on a zig-zaggy route from Port Townsend, Washington, in the north; to Bakersfield, California, in the south; to Austin, Nevada, in the east; and to—yes—Orick, California, in the west. I had researched the story of the giant peanut there, so we made a point of locating it when we passed through.

The sculpture, despite its size and besides its deterioration, is uninspiring. It embodies a short-sighted viewpoint one could call, "Jobs *über alles*": jobs over trees, jobs over rivers, jobs over animals, jobs over human health.

Jobs over *everything*, really. This viewpoint reigns supreme all along the conventional political spectrum in the USA, from right to left, from libertarian to liberal, from CEO to worker. Only a handful of radicals try to stand outside and take in the wider perspective. Over the years, many well-meaning thinkers and activists have genuinely sought to create a magic balance between jobs and the environment (and many political and corporate entities have claimed to provide one) but that hoped-for state has remained elusive because—as our travels showed us—it is an impossible dream. Capitalism's imperative for "growth" is antithetical to the environment's essential health.

Everywhere we journeyed in 2014, we saw the ravaged landscapes left behind by the pursuit of Jobs: clear-cut forests, drained wetlands, dammed rivers, trampled deserts and mined hills. But this destruction is not the whole story. What we also saw everywhere—during what turned out to be the hottest summer on record up to that point—was something global in scope: Climate Change.

THE FORESTS

From the Olympic Peninsula in Washington to the southern Sierras in California, the forests of the Western USA have been relentlessly clear-cut, in many places repeatedly. Of the old growth that existed in these areas before the European invasion, less than 10% remains. For particular species, the number is even worse: Over 96% of the coastal Redwoods that existed in the mid-1800's are gone. As if this isn't bad enough, the logging of old growth is not a legacy of the past; it continues to this day. The sections that remain are besieged islands, surrounded by plantations. Tree farms takes up most forested landscapes from horizon to horizon.

Clarabelle and I camped in a patch of old growth near Oregon's Mt. Hood which I had first visited in 2002 when it was on the chopping block as a proposed timber sale. Forest defenders had built a tree-sit in one of the Douglas-Firs, 110 feet off the ground, and this action prevented logging while a legal battle was waged. In a story that's far too rare, the activists eventually won and the timber sale was canceled. The spot is one of my favorite places in the world. Just standing among a group of such big, tall, old trees is awe-inspiring. Because this particular site is off the beaten path and not marked on any tourist maps, it also has solitude.

But the entire canceled timber sale was small: only 167 acres, broken into eighteen mostly non-contiguous units. That's just a quarter of a square mile. You can't even get lost in such small pieces of forest. Just outside their well-defined boundaries are stumps and tiny trees. Though ac-

2

tive logging has not happened in the area for at least 20 years, the areas of old growth are constantly decreasing in size anyway; because the edges of the groves are over-exposed, the wind takes out trees over time. In the decade I have been visiting, the ground in some areas has gone from mostly open to being an impenetrable clutter of enormous logs.

Driving through the forests of the West, most tourists don't see past the "beauty strip," also known as the "idiot strip." This is the buffer of trees along the road that screens the clear-cuts from immediate view. But Clarabelle and I were looking more carefully and we couldn't miss what was really going on: a vast patchwork of tree plantations, laid out with straight-lines and right-angles regardless of terrain. In each section, all the trees were of uniform size and shape, in contrast to natural forests where a mix of ages are present. Forty foot trees will be followed by saplings and then twenty-footers, each in their own well-defined parcel like fields of corn, alfalfa and soy. You don't have to visit the West to see how bad it is; look at satellite photos on your favorite mapping application and the dividing and conquering is clearly visible.

This is "forest management" as prescribed by the National Forest Service, which—as most people don't know—is part of the US Department of Agriculture. The trees are being farmed the way any other crop is: by wiping out the native ecosystem with machinery and chemicals, breaking it up into countless individual parcels, and maintaining that state of fragmentation through continual disruption.

THE FIELDS

Most people don't think of Agriculture as being destructive to the environment. After all, compared to the concrete and glass jungle of the city, farm fields are green and open and look, well, "natural." But this is hardly the case. Our trip took us through California's Central Valley, from Red Bluff to Bakersfield, and frankly, I was shocked. The completeness with which the native landscape has been removed and is intensely managed for farming —leaving no room for anything else—was not something I had expected.

The orchards in particular were an impressive sight. On either side of the I-5 or the 99 (both of which run north-south through the Valley), ruler-straight rows of trees stretched into the distance, as far as the eye could see. Every tree was pruned to the same shape, to better allow machine-assisted harvesting. The ground below was absolutely clear of any vegetation. The only way to keep out weeds like that in an orchard is with herbicides; the trees' roots are too close to the surface to allow tilling under unwanted vegetation. This dead-zone under the trees extended to the fence along the freeway and on the smaller roads it came right up to the shoulder.

The fields of grape vines and row crops were kept up the same way, all in huge sterile blocks. For long stretches of highway, we would see only one species of plant at a time. We wondered what the native flora of the area might be but we had few clues. Could there be more diversity of plant life growing out of the cracks in the sidewalk in the city of Sacramento than in the fields of the Sacramento Valley? Maybe.

But it's not just plants that have been eradicated from farmland; animals, too, are victims. Where are the herds of ungulates who grazed when these field were grasslands? The predators who hunted them? The birds and amphibians who thrived in the wetlands? The butterflies and insects? Rachel Carson warned us over fifty years ago that we were risking a "silent spring." Now, a greater dearth of wild voices is nearly upon us, replaced by the growl of machinery, in a landscape nearly empty of wildlife. But the tourist exclaims, "I love driving out to wine country on the weekends!"

One could respond that, "We have to eat, don't we?" and of course we do, but this style of agriculture is not the only one available to us. A study sponsored by the U.N. found that small-scale agriculture of mixed crops using no pesticides produces more food, acre-for-acre, than large-scale monoculture.[1] Historically, the farming methods used by the original human inhabitants of the Americas were far less destructive and successfully fed many millions of people. (For a good survey of some of these, see Charles C. Mann's book, *1491: New Revelations of the Americas before Columbus*.)

Big corporate agriculture as exemplified in the Sacramento Valley (and in Oregon's Willamette Valley, and across the entire Midwest), is used so extensively not because it is the best way to feed people but because it is the most profitable in the short-term. Not figured in are the costs of the long-term effects: topsoil degradation, aquifer depletion, air pollution, species extinction and Climate Change. Mainstream economists refers to such factors as "externalities." "'External' to what?" one might rightly ask. Only to the standard theoretical models. They are not external to the thing called "the Real World."

THE MOUNTAINS

In Kernville, near Bakersfield in southern California, we visited Clarabelle's brother, a wildland firefighter stationed there that season. He and other military vets on the crew referred to the area as "Little Afghanistan" because the scrubby, treeless hills had a similar appearance.

We drove up into those hills looking for an out-of-the-way place to camp and found one at the opening of a water-cut ravine on a spur road at about 6000' elevation. A small but vigorous stream flowed noisily over smooth rocks, under fallen logs and among vehicle-sized boulders, alternately cascading down waterfalls and collecting in pools. A few trees, mostly pines and a few cedars, huddled at the mouth of the ravine, sheltering a small sandy bank. Quite the lovely spot, but marred by one thing: cow manure. There were patties all along the bank and in the watercourse itself. So we couldn't drink from that stream, and even filtering it seemed sketchy.

Most people in this day and age have never had the experience of drinking directly from a stream, so they don't know what they're missing. It is not merely a pleasure, but also a re-enactment of a time when we all lived direct-from-the-source, in intimate inter-connection with our environment. Today, our relationship with nature is us-vs.-them. Always drinking treated water that is piped or bottled must affect us, and not for the better. I can tell when I'm kneeling at a stream bank, bringing my cupped hands to my mouth, that the liquid is not just different for lacking chlorine; more than my physical thirst is quenched when I drink it.

But you can't drink from a watercourse that's got cow shit it in, and that's what you'll find across most of the West, where ranching rules. Cattle are officially grazed on 70% of public lands in the seven western-most states, though the actual number is higher given that illegal grazing is rampant. Cows tend to gravitate to the wettest places they can find, but in these arid places, riparian zones (areas with running water and wetlands) are a very small part of the landscape, making up less than 3% of the Great Basin, for example. As a result, the damage disproportionally affects places that are rare as well as delicate. Cows and sheep are quite harmful because their behavior is alien; no analogous species existed previous to their introduction, so the flora and fauna are not adapted to deal with them. Habitat destruction results, threatening many native species, including traditional food plants of the Indians.

Ranching and the wild don't mix: fences disrupt migration patterns of wild animals; native predators such as the wolf are hunted nearly to extinction; water is diverted (and often wasted) leaving less for creatures who depend on it. These effects and more are happening over literally millions of acres, but the damage is invisible to most people. Who knows what a healthy sage-brush steppe ecosystem looks like anymore? Ungrazed examples are as rare as old-growth.

CLIMATE CHANGE

Scientists say it's tricky to attribute a particular weather event like a drought or "super storm" to Climate Change. But an excellent way of understanding the issue appeared in a 2012 article in the *Bulletin of the American Meteorological Society*, and has since become well-known: The Baseball Player on Steroids analogy.[2] Imagine that a baseball player starts taking steroids and hits 20% more home runs that season than the previous one. While it might be impossible to know if any one particular home run can be attributed to his steroid use, the *probability* of his hitting home runs has increased by 20%. Climate Change works in a similar way. The global rise in temperature is analogous to the baseball player's steroid use and has increased the probability of extreme weather events. Though attributing particular events to Climate Change might be impossible, their probability has increased.

It is with that caveat that I mention some of the extremes and abnormalities that we witnessed in 2014:

- Smoke was omnipresent in the skies of southern Oregon, from Ashland to Grants Pass to Cave Junction and all places between, high and low, all summer long. Sunsets were stained red, moonrises orange. The "Fire Danger" signs posted at the entrances of the National Forests were all set to "EXTREME."

- I started this essay in October while sitting on the sandy shore of the Kern River in southern California. The Kern is well-known as a whitewater rafting destination but after a decade of drought the flow was too low. A local camper familiar with the area told us our tent-site near the riverside had always been underwater in previous years.

- In the same area, bear sightings close to town and along the roads have increased. Searching for food, they are coming down from the drought-stricken hills. This behavior puts them in increased danger of being hit by cars or attacked by humans.

- The city water in Harbor, Oregon, was unsafe to drink when we visited in September because of salt contamination. With the lack of rain, the level of the river had dropped and become inundated by the sea at its lower reaches where the intake pipes were located. Water trucks were providing free fill-ups for people who brought their own containers. Moving the intake pipes would be a costly operation and the community hoped to avoid that.

- Nearly every lake we saw was low, and some, like Lake Shasta, were drastically so. We observed more boat docks stranded on dry land than standing in the water.

- Camping at Lake Pillsbury in October, we were entranced by the eerie vocalizations of the Elk. The bellows of the bulls and the squeaks of the cows sound almost like whale song. But the park ranger told us it was unheard of for the bulls to still be in the area that late in the year.

Fragmented forests on fire from a climate over-heated by cutting too many trees; rivers and lakes drying up; animals changing their habits; whole ecosystems withering and losing their resiliency: Climate Change is here, but not many people seem to be noticing it yet.

GOVERNMENT, MEDIA & LIES

We picked up the local newspapers as we passed through human habitations between camping spots. The Internet Age has not made newspapers entirely irrelevant yet, especially in small cities and rural areas, and we were able to glean some flavor of the character and priorities of a locale from these publications. Every paper in Oregon and northern California was prominently covering the summer's wildfires, usually front-page and above the fold, enumerating the acres engulfed, total structures burned and percentage contained. In grocery check-out lines and at gas stations, the fires and the weather were the prime topics of conversation (usually with a wish for rain).

But only rarely, in print or in person, did anyone make the connection between the fires and Climate Change. Variations on the same old "jobs" refrain were much more common in the media. Consider this pair of articles from the Sept. 20th edition of Curry [County, Oregon] *Coastal Pilot*, both penned by staff writer, Jane Stebbins:

Senate to decide forest bill

For the third time, the US House of Representatives has approved a forestry bill penned by Rep Greg Walden (R), crafted to create jobs in the woods, improve forest health, reduce the risk of wildfire and generate funds for local communities....

[Claimed Walden:] "The legislation in this package... would allow us to put people back to work in the woods, reduce fires and produce revenue for schools, teachers, sheriffs and sheriffs' deputies, search and rescue—for all these basic services that matter in rural communities across the West."

...HR 1526 could generate as much as $90 million a year for struggling rural Oregon counties by reopening the forest to logging....

"You want to do something about poverty? Create a job!" Walden told the House. "You want to get America back on track? ...We'll create jobs, generate revenue, and have positive cash flow in this country for once. It doesn't have to be this way." [my emphasis]

Is Walden lying or is he sincerely misled? I don't know, but his final words are definitely true. It *doesn't* have to be this way. And if Oregonians gave an honest look at their state, they would see that it's no longer rich in resources.

Stebbins had a related article on the same page of that day's paper:

County eyeing national lands issue

Curry County commissioners are keeping an eye on Arizona and Utah, where legislators are trying to get the federal government to relinquish national lands to the states for use as they see fit.

The issue has been increasingly in the news since the national government has ended federal timber payments to half of Oregon's counties and other counties throughout the West realize the vast majority of the lands aren't paying for themselves [my emphasis].

...[In Utah], the BLM is set to auction off 27 parcels totaling 29,400 acres near the Green and White rivers for oil and gas leasing. Recently, the Southern Utah Wilderness Alliance filed an official protest in an effort to halt the lease operation on grounds that the land possesses wilderness qualities.

But locking up more land as protected wilderness means less control the state has over those lands—and less money it might be able to garner otherwise...

In its heyday, Curry County received more than $6 million from the federal government—funds it paid in lieu of a percentage of tax revenue the county would have garnered had environmentalists not shut down the forests to timber harvests in the 1990s.

The veracity of Stebbins' characterization of the 1990s as a time when the forests were "shut down" to timber harvests is contrary to the facts. Clinton's Northwest Forest Plan and the notorious "Salvage Rider" put thousands of acres of old growth trees on the chopping block, surpassing the amounts of his Republican predecessors, Reagan and Bush I. Naturally, these policies provoked an outburst of eco-defense activities, including tree-sitting, but for every place that was saved, many were not.

More to the point is the fact that we can no longer afford the ridiculous viewpoint that the Earth needs to "pay for itself"; that it is a collection of resources from which to "garner" (as Stebbins said twice) money. Such a viewpoint is only too obviously rooted in the "dominion" over all living things granted to Adam, and hence all of humanity in the Book of Genesis. Contemporary adherents of the Abrahamic religions who revere that book —Jews, Christians and Muslims alike—need to recast their dogmas and help stop ecological destruction. If they don't, all their temples, churches and mosques will stand empty on a planet that's become inhospitable for human habitation.

Is such a possibility a shrill, Chicken Little exaggeration? Not at all. A report by the Global Carbon Project demonstrated that carbon emissions in 2013 continued to increase at a "worst-case scenario" rate. This is a rate of emissions that will have devastating effects to life on the planet according to the UN's International Panel on Climate Change (IPCC), which is considered overly conservative to many in Climate Change circles. As stated on the long-running climate blog, "robertscribbler": "At the current pace of emission, it will take less than 30 years to lock in a 550ppm CO_2 equivalent value—enough to melt all the ice on Earth and raise temperatures by between 5 and 6 degrees Celsius [9-10.8 degrees F] long term."[3]

That's a planet that humans can not survive on.

Not just small-town newspapers are missing the point, of course. Nearly every outlet in the entire US American media establishment is guilty, almost all of the time. In Ashland, we picked up a copy of the San Francisco *Chronicle*, whose Sept. 24th edition had a front-page article discussing "renewable" energy:

State's deserts seen as sites for huge new plants

Industrial-scale solar, wind and geothermal projects could be built within a few miles of national parks in the California desert as part of the Obama and Brown administrations' efforts to combat Climate Change, under a mammoth plan released by federal and state officials Tuesday.

Construction of the plants, many of which could cover several square miles, would drastically alter desert vistas near national parks and wilderness areas, according to a draft [of the plan].... But that would be offset by the climate-change benefits of allowing large solar and wind energy plants on more than 2 million acres of the Mojave Desert, the report said.

Energy plants covering several square miles are obviously going to affect more than "vistas." Wind farms are notorious for killing birds by the thousands. Construction of anything that large is going to make mincemeat of the creatures living there, whether plant, animal, or other. The same article mentions the 5.4 square mile Ivanpah solar plant in the Mojave Desert, where "[m]any more desert tortoises were discovered on the site than were anticipated, and problems have emerged with birds being incinerated when they fly in the path of the concentrated solar radiation." Birds being *incinerated*? That's horrific! This is how we are going to mitigate Climate Change? By running heedless over any and every other living thing?

The article quoted one person who is calling bullshit on the plan:

Kevin Emmerich, a former Death Valley park ranger who founded the watchdog site Basin and Range Watch, said officials are "calling this a conservation plan while they are planning on fragmenting up the large remaining sections of the California desert into 'development zones,' which ultimately translates to a net loss of desert habitat."

Emmerich said planners have refused to consider roof-top solar and other smaller-scale projects that "would actually produce energy at the point of use without transmission loss and save habitat."

Kudos to Emmerich! However, the SF *Chronicle* article ends with a fundamentally dangerous lie:

But Mark Tholke, a vice president at EDF Renewables, an energy company that has built plants in the desert, said rooftop solar is inadequate to address Climate Change.

"Many of us feel a real urgency to get as many (plants) up and running as possible, as soon as possible," Tholke said. To slow Climate Change, he said, "we need to do a lot more than rooftop and distributed generation. We need cost-effective, large projects."

False. In actuality, what "we need to do" to slow Climate Change is a lot less of *everything*. We must re-localize our means of living, not just for energy, but also food, shelter and medicine. That's a huge project and is enough to keep everyone busy.

WE MUST CHOOSE THE ENVIRONMENT OVER JOBS

The days of the massive power plant, the 2500 mile salad, the factory-produced pharmaceutical, and the car-centric megalopolis must end. So, too, must go that holiest of holy grails: "Jobs." Jobs provide people with money; the more money circulates, the more the economy grows; the more the economy grows, the more destruction is meted out on the Earth. Yet, favoring a raise in the minimum wage is considered "progressive." Typical was this article, from the *Mail Tribune* ("Southern Oregon's News Source," based in Jackson County), Sept. 22, reprinted from Dallas *Morning News* under heading of "Other Views":

Congress should pass wage hike

A vibrant economy requires consumers who are able to earn enough to provide for their families as well as employers who are able to make a profit... The Economic Policy Institute estimates that phasing in a federal minimum wage hike would bolster the income of 27.8 million workers, expand GDP by about $22 billion, add 85,000 jobs and most benefit adults struggling in $8 and $9 hourly jobs, who also *would probably receive a wage hike. About 75% of Americans, including 58% of Republicans, support increasing the federal minimum wage.*

The folks in Orick had it wrong in the Carter administration, and supporters of a higher minimum wage have it wrong now too. We can't *even* afford to "work for peanuts"—*that's* paying too much. Instead, business-as-usual, techno-industrial society as we know it must be brought to a screeching halt. No more growth. No more debt. No more jobs. It's not that we don't make enough money; it's that life is monetized and it shouldn't be.

The cliché threat of the mugger sums up perfectly the choice before us in these times: "Your money or your life."

March 16, 2015

NOTES:

[1]   Meyer, Nick. "UN Report Says Small-Scale Organic Farming Only Way to Feed the World" Technologywater, Dec. 14, 2013.
      https://www.technologywater.com/post/69995394390/un-report-says-small-scale-organic-farming-only
[2]   Peterson, Thomas C. "Explaining Extreme Events of 2011 from a Climate Perspective" Bulletin of the American Meteorological Society, July 2012.
      https://doi.org/10.1175/BAMS-D-12-00021.1
[3]   Fanney, Robert Marston. "Worst Case Carbon Dioxide Emissions Increases Continue — Hitting 40 Billion Tons Per Year in 2013" robertscribbler blog, 9/22/2014. https://robertscribbler.com/2014/09/22/worst-case-carbon-dioxide-emissions-increases-continue-hitting-40-billion-tons-per-year-in-2013/

Taking on the Sacred Cow of Big "Green" Energy—Interview with Basin & Range Watch

The deserts of the American Southwest have come under a new assault in the last decade. The few, fragmented areas of these austere, rugged, yet delicate landscapes which had managed to survive relatively intact from mining, ranching, military use (including nuclear tests), urban encroachment and motorized recreation, are now being targeted for large-scale "green" energy projects, many of them on public lands.

Early during Obama's first term, his administration offered a wide array of federal incentives including grants, loan guarantees and tax breaks for renewable energy projects, ostensibly to reduce the nation's carbon footprint. Mainstream environmentalists greeted this with cheers, but as was characteristic of Obama's administration, the hope turned out to be hype. Big corporations have been the beneficiaries of his programs and the environment is still the big loser.

Basin & Range Watch is a non-profit that operates out of Beatty, Nevada, in the Mojave Desert. Their mission is to "conserve the deserts of Nevada and California and to educate the public about the diversity of life, culture, and history of the ecosystems and wild lands." Central to this mission is opposing the many large-scale solar and wind projects that have been proposed in the area, a number of which have been built with deleterious consequences. In these efforts, Basin & Range Watch has found itself at odds not just with big corporations and big government but also with the big environmental organizations (who depend on funding from big non-profits). But someone must speak for the deserts!

I first heard of Basin & Range Watch in the autumn of 2014, when co-founder, Kevin Emmerich was quoted in a San Francisco *Chronicle* article about siting renewable energy in the desert. (See "Your Money or Your Life: Can We Afford to Work for Peanuts?" in this volume.) Through their website I learned about the threats to wildlife habitat, endangered species, aquifers, recreation areas and scenic vistas, as well as to small-scale renewable energy projects at the community level. So when I found myself in their corner of the desert in 2016, I contacted them.

On the drive to Beatty from the Mojave National Preserve, where I'd been camping, I found myself at a high point along Interstate 15 with a spectacular view of the Ivanpah Solar Electric Generating System. Ivanpah is a very large-scale example of what is called a "concentrated thermal solar plant." It functions by using mirrors—in this case, 173,500 *pairs* of them —to focus sunlight on central towers that contain water boilers. The resulting steam turns turbines that generate electricity.

The system is not carbon-free in its daily operation, requiring natural gas to start it up every morning. According to Wikipedia:

> *In 2014, the plant burned 867,740 million BTU of natural gas emitting 46,084 metric tons of carbon dioxide, which is nearly twice the pollution threshold at which power plants and factories in California are required to participate in the state's cap and trade program to reduce carbon emissions. If that gas had been used in a conventional fossil fuel plant, it would have generated nearly 124,000 MW·h of electrical energy. That is enough to power the annual needs of 20,660 Southern California homes.*

Ivanpah is notorious for regularly killing birds by literally burning them—sometimes to the point of complete incineration—when they fly through the super-heated air around the tower, which can reach temperatures of 1000°F. It's brutal. So is the footprint of the power plant in the desert where thousands of animals and innumerable plants once had their homes. I was quite taken aback by its scale. I'd seen photos but they don't do it justice.

Ivanpah has also been plagued with performance and financial issues. It has never produced as many megawatts as promised, its power is expensive to purchase, and the entire project is in danger of defaulting. The word that comes to mind is, "boondoggle."

The day after I saw Ivanpah, I interviewed the founders of Basin & Range Watch, Kevin Emmerich and Laura Cunningham, at their home and headquarters. What follows is a transcript of that conversation, edited for length and clarity.

SONNENBLUME: How and why did you start Basin & Range Watch?

EMMERICH: We lived here for a little while and we decided that no one was giving a lot of attention to some of these flatlands in Nevada, some of these unknown mountain ranges, some these unknown areas. We decided we wanted to create an educational type of website that would talk about these areas and some of the issues that face them. We got into mining issues and an off-road race, which was quite controversial. In about 2008 or so was the land rush of big solar and wind energy. The current administration [Obama's] really wanted a lot renewable energy and a lot of it on public lands and opened almost everything up to big solar and wind applications. We looked at map of the Amargosa Valley, and it was covered in applications and we really wanted to make this an issue. Other people wanted to but it was "green" big energy, and because it was "green" big energy, it was a sacred cow and it's something that a lot of environmental groups and environmental reporters really didn't want to touch. We've been doing it for a while, like seven years now.

CUNNINGHAM: I went to Berkeley and studied biology there and then through the years got into the field of tortoise biology. I went into the Mojave Desert and worked as a contract biologist for these companies where you literally march along with a shovel—all this university education and it came down to carrying a shovel!—walking within survey lines, ten hours a day, digging out tortoise burrows, bashing creosote bushes down to the ground, to get every last tortoise out of a 20 square mile area before the bulldozers come for large projects. I learned what happens when you mitigate for tortoises. It is uglier than people think. You physically destroy desert ecosystems, and usually 50% of the tortoises die during that translocation process from stress or predation. So that's when I joined Kevin. (He's my husband, too.) We saw the same things are going to happen when you build a giant solar project on 20 square miles or 10 square miles. You're going to have to hire the tortoise biologists and dig tortoises out, and then huge earth moving machines come and just destroy this beautiful wildflower desert ecosystem.

We had a big push to educate people that there's got to be better alternatives, because these desert soils store carbon, for one thing, and when you bulldoze them up it releases carbon. And all the roots and plants and mycorrhiza, and really interesting little puffball mushrooms that come up in a spring like this [when there has been enough rain]: all this biodiversity that we're destroying when we do large-scale—even renewable—energy in these deserts.

SONNENBLUME: And also, what are they called—"cryptobiotic crusts"?

CUNNINGHAM: Exactly! I have pictures of mosses, and even little liverworts and blue-green algae, and you can see them in little patches here and there, and they'll green up in the winter rains. I asked, how are you going to mitigate that? They said, "We might be able to roll them up like a lawn."

SONNENBLUME: [*laughing*] Like sod?

CUNNINGHAM: [*laughing*] Right. Like sod. Then they can put them somewhere else. But they didn't even try because they knew that wouldn't work.

SONNENBLUME: I think there is a misunderstanding that a lot of people have about the desert where they think, "Oh it's empty, there's nothing out there." People don't see the amazing life.

CUNNINGHAM: Exactly. So we would always try to write comments and say there's better alternatives. Maybe not the giant power towers, but definitely the solar photovoltaic panels can go on parking lot rooftop structures, even small disturbed lots in urban areas. So we've been trying to shift alternatives to the built environment, but that's been a really tough sell.

EMMERICH: What's fascinating about that perception of the desert as a wasteland is the way it's morphed over the past six to seven years. Whereas in 2008 [when the Ivanpah solar array project was started], they were saying that the Ivanpah Valley was not worth that much, because of the power line and the highway and a casino an golf course. But these are little dots compared to the amount that they developed. They portrayed the site as having lower biodiversity because if you go to the other side of Ivanpah Valley there were more desert tortoises. So they had to really to spin that "wasteland" portrayal to support a lot of these large-scale energy sites as really "worthless."

But now we've got Senator Harry Reid designating the Basin and Range National Monument in Nevada. It's really a beautiful place, but there isn't a basin in Nevada that's not beautiful! It looks exactly like the five solar energy areas that were called visually "worthless, ugly." There's one up the road that I think is as beautiful as anything in the monument. We're able to point out some of this hypocrisy now. You really can't find a big piece of desert to put a giant solar farm in that's not going to be a visual clusterfuck.

SONNENBLUME: And that's going to affect a whole ecosystem.

EMMERICH: Exactly. That's true with the Ivanpah Valley. They said it was only going to be 25 desert tortoises because that's what the biologists found in a survey. But really it turned out to be like a 150.

CUNNINGHAM: When they went in with the shovels and dug all the burrows, the number more than quadrupled because tortoises spend most of their life underground.

SONNENBLUME: Like, over 90% of the time.

CUNNINGHAM: Right. It was the mayor of Barstow who said at a public meeting [about a proposed large-scale renewable energy project]: "We get Chinese tourists in big buses. They come from China and get off the bus and they say, 'There's nothing here!' They value that because they don't have that where they're from." They literally come to see the nothingness of the great desert.

EMMERICH: The perception of the desert changed when people saw what happened at Ivanpah.

SONNENBLUME: I saw it yesterday. It's astounding. Obviously the fossil fuel burning is a problem—something else needs to happen—and so there's the interest in wind and solar because they are, quote, carbon-free. Although they're not in the sense that there's a lot of carbon pollution involved in their production and then there's the extraction of all the rare earth metals that are needed. Maybe you could talk about the corporate connection—making money off these things—and how people are not looking at the option of localized solutions.

EMMERICH: First off you have to ask yourself, how many megawatts do some of these really big wind farms produce? Ocotillo, for example. They said it would produce power for 130,000 homes but it's only running at maybe 50% of its capacity. So how much is that really replacing? As far as the last part of your question, the alternative is that we can easily get that amount of megawatts through using the developed environment. There's so much room in Vegas for example, so much that's wasted down there.

There's a big corporate tie to owning energy. And there's even a lot of solar and wind projects that are being built by British Petroleum and a lot of people see that as a good thing because here are these fossil fuel giants moving into renewables and big solar. We've also got Warren Buffet getting big tax credits for building big projects. That's where the corporate ties are. There's huge tax credits for really big companies that are more interested in producing energy and making a lot of money [rather] than being "green."

CUNNINGHAM: It's not like this is us little people controlling this. We've been looking at what we call "energy democracy" on one side and the corporate ownership of renewables on the other side. You get a lot of tax credits [as a big corporation]. In the past there have been federal grants and Department of Energy loan guarantees to really help build these projects such as Ivanpah. So there's been a huge push of taxpayer money to build the central station utility plants. But on the other hand, the more distributed generation types of solar that individual homeowners can participate in have been tamped down. Even in Nevada, there's a raging controversy right now because a lot of people want to participate in a net metering system where you get a little incentive from the utility. It's profitable for them [the utilities] because they will purchase your excess energy and sell it on the grid. It's not like they are losing money on it. The home owner has the incentive to help pay for the rooftop solar, which is good for everybody, good for Climate Change. The Ivanpah project got a billion dollars in incentives. A grant. Free money to build the project. [Actually, it was $1.6 billion! -KtS]

But the utilities have really been pushing back on this energy democracy model and lobbying politicians and they successfully reduced the incentives. And they're not raising the caps. There's actually caps for how many houses can have rooftop solar [and be eligible for the incentive]. If Climate Changes is an emergency, you'd think we would want to have no caps—get this as extensive as we can.

SONNENBLUME: There's caps?

CUNNINGHAM: There are caps. In Nevada and in California. I was just reading this morning that PG&E [Pacific Gas & Electric] is approaching the cap for how any houses in the PG&E utility district can participate in net metering. It's so popular. The cost of photovoltaic panels has come down so much. People are pushing to put rooftop solar panels on their houses, but there's a cap and we're going to reach it. Then there will have to be public utility commission meetings about whether we should raise the cap.

There is a big push back from the utilities and it's not solved yet because there are benefits that the homeowners are providing to the grid that the utility companies benefit from, so it benefits everyone, not just the people with rooftop solar. For example, the peak time of use is right when all those rooftop systems are feeding electrons back into the grid when everyone's air conditioning is running. There's arguments that utilities won't have to construct so many natural gas "peaker" plants to keep up with the peak time of day. It helps stabilize the grid having these distributed systems everywhere.

It parallels local, organic, urban gardening: Small systems that are infilling in urban areas. You don't have to truck produce and vegetables miles and miles. Rooftop solar is the same thing. You don't have to build a 500 kilowatt transmission line 300 miles long. That Ivanpah transmission line to L.A.—they had to upgrade it and all rate payers had to pay for that. So there's an interesting parallel between local food production and local energy production. We're trying to explore this but it's hard when you have the big utility companies pushing back on you.

And people don't realize the facts about these transmission lines. When I was a biologist, I took a safety class because I was going to work for one of the utilities. They built a 500 kilovolt transmission lines—giant metal towers—to get the wind power from Tehachapi Pass to L.A. People don't realize what that takes. I was amazed at the amount of impact that has. They had to build it through this beautiful Joshua Tree forest in the West Mojave Desert and they just took out hundreds of trees. I had a friend who worked on it. There were Joshua Trees she couldn't put her arms around, their trunks we're so big, and they knocked 'em over, killed them, to get these towers through. They tried to salvage some but they mostly didn't. Then they needed to get this transmission line over the San Gabriel Mountains which is a huge, rugged, steep, rough mountain range and that involved helicopters putting in towers and blasting the sharper ridges. One of the biologists died because he fell off a cliff while he was monitoring the impact. So people were dying building these renewable projects.

There's a human impact. It had to go through the city of Chino and the whole city of Chino said, no, we don't want this giant transmission line plowing right through our neighborhoods, so the city of Chino sued Southern California Electric and they lost. They used eminent domain through Chino. There's quite a lot of impacts that people aren't aware of for green energy.

SONNENBLUME: The transmission lines lose energy along the way, too, right?

CUNNINGHAM: Yes, about 10%.

SONNENBLUME: So everything that leaves Ivanpah doesn't arrive in L.A. When it's coming from your own roof, you're losing essentially nothing.

CUNNINGHAM: That's right.

EMMERICH: And Ivanpah produces so little, you don't want to lose 10%.

CUNNINGHAM: Yup, they've been having trouble. What you're seeing in that valley is one giant concentrated solar plant and there's two photovoltaic plants next to it. When clouds come over—even just one little cloud over the sun—it shuts the whole plant down.

So that's a benefit of distributed generation, all the solar rooftops in the city interconnected with microgrids on the whole grid: it dissipates that cloud cover loss. Obviously there are still things that need to be worked out. We get a lot of push-back from people saying it's too hard to do rooftop solar. And there are problems but when we look at these large scale projects out there, they have enormous problems to overcome. Ivanpah has never been at 100%.

SONNENBLUME: Is that typical of these large projects? Solar and wind both?

EMMERICH: It's not been uncommon for the concentrated solar thermal plants [like Ivanpah] and the Solaris plant in Arizona. With photovoltaic, the jury's out. They say they "overbuild" them so they can get more megawatts on cloudy days but I just don't know. They have a peak time of day when they can sell energy and everyone's competing for that.

CUNNINGHAM: The wind projects, too, have often been built in low wind resource areas and they're running very low. The siting for large wind and solar has often been as close to existing transmission lines as possible, not [asking] are there tortoises here, or bats or golden eagles or even wind! It's often been just ease of transmission.

EMMERICH: A lot of this projects have facilitated the need to build base-load natural gas plants, like Sentinel, near Desert Hot Springs. It's used as a back-up for the big renewable projects that were built to the east. Natural gas plants don't have as big of a land footprint but they have a big smog footprint.

The big thing is going to be energy storage. Energy storage for renewables is a long ways away. The best one so far is hydro-electric but they don't really have what I would call a holy grail. They have this plant north of us called Crescent Dunes and it's [a tower] like the Ivanpah power plant but they have storage in it, [in the form of] molten salt. And that's a big deal to be able to do that but they needed a $730 million loan, and they were over-budget and it took ten years to complete it. They're not going to be able to build a million of these molten salt towers worldwide without funding. That's the future: where is the storage of the energy?

SONNENBLUME: Right, because people who have solar panels on their cabin or wherever have batteries they charge up, and it's like that on a larger scale, too.

CUNNINGHAM: Right.

SONNENBLUME: There's a storage project I read about on your website planned near Joshua Tree National Park, in the final stages of approval. I can't remember what it's called.

CUNNINGHAM: Eagle Mountain. It's ridiculous. It's a giant old mine pit and they want to fill it and make it a huge reservoir and then they would drill these giant tunnels through the rock and they would pump water using excess wind and solar from the large scale projects into the upper pool. Then they would let it go through tunnels to turn turbines, like a dam. But we're like, where are they going to get all this water? This is an enormous reservoir in the middle of the desert and there's barely enough water for farms and residences down there.

SONNENBLUME: The National Park Service is opposed to it.

CUNNINGHAM: Completely opposed to it.

EMMERICH: It''s supposed to be fossil water.

CUNNINGHAM: Right. Ice Age water.

EMMERICH: They will lose hundreds of acre feet a year—

CUNNINGHAM:—through evaporation—

EMMERICH:—and seepage.

SONNENBLUME: There's several dozen oases left in Joshua Tree National Park and they're getting their water from fissures in the crust where this ancient water comes up, and that's why they are there. And these oases are fascinating places because that was the dominant ecosystem millions of years ago. These are little relic areas that have managed to survive all this time and now that's the threat to them. This is the concern of the Park Service, in part, that a falling water table in the area will kill them off.

CUNNINGHAM: For a very small amount of storage capacity.

EMMERICH: That one has a good amount of backup capacity. [The issue is] the amount of water it will use. But the question is, do they even have a need for it? Have they even figured out who will use it? Nobody has.

SONNENBLUME: Isn't there a perversity where not all of these projects are actually responding to a specific demand? Some of them are occurring because there's money in building them?

CUNNINGHAM: Yes. There used to be these American [Recovery and] Reinvestment [Act] era grants and they would literally give hundreds of millions to these projects and we think it forced them to build too big. Ivanpah has had a huge number of problems because it was scaled so gigantically because money was funneled into it to make it gigantic. You can build really small power towers that are 50 feet tall and they could be distributed in empty lots in cities. Same technology. But that wasn't charismatic enough for politicians. They wanted something really big and showy. There's been big problems and it's been expensive to solve them. So we're always looking for small and in the built-environment. And energy efficiency. We should be doing energy efficiency before any of this but that wasn't on the radar.

SONNENBLUME: Like insulation, double-paned windows: the basics.

CUNNINGHAM: Yes, the basics. But it's been backwards. You have these huge centralized power plants that have been really slow to produce but they look good and everyone's impressed. The glittering mirrors dazzle reporters. Then it's been very slow on the side where homes are made more energy efficient and we improve our grid in the city to be able to handle more renewable energy.

SONNENBLUME: How much would they have to do with solar and wind to replace a significant portion of coal, for example. What would that look like?

CUNNINGHAM: I've been studying the possibility of 100% renewables and I don't think we're capable of that. Germany thinks it could get to 60% or 80% and they're not sure how they would get to 100%. Again, we need to reduce our usage of electricity, which we're not really talking about. Our high standard of living is a real obstacle. Kevin and I talk to engineers and [they] are always telling us, this is going to be a very heavy lift if we're going to keep our very high standard of living.

EMMERICH: Aspen, Colorado, gets something like 75% of their energy from renewables, but that's because they have a dam right there.

SONNENBLUME: And dams have their own issues.

CUNNINGHAM: We totally don't support more dams blocking salmon. Salmon are almost extinct in California so the last thing that's green is to dam more rivers. We don't want to do that. Hydro isn't exactly the greenest form of energy but that might be the only way to get to 100% because you can control how much water is let out and that will be a big balance to the intermittent peaks for wind and solar.

EMMERICH: What about the carbon footprint [of the big renewable projects]? Like Blythe Solar Energy Project. That was going to be a horizontal parabolic trough plant with mirrors and there were going to be four units and they were bragging how this plant is going to use as much steel as it took to build the Golden Gate Bridge. And I'm like, whoah, that's one hell of a carbon footprint. But we're going to save the planet from global warming.

We were morbidly fascinated at the construction of the Ivanpah project and I hung out there a couple times at quitting time and there were 200 cars leaving work and some were going to Barstow [over 100 miles away].

CUNNINGHAM: We're trying to study life-cycle impacts of these projects. Even rooftop solar. There are impacts to any energy used. Battery mines. The lithium mines north of here, near to the town of Silver Peak which—

SONNENBLUME:—this is the one that Elon Musk and Tesla is involved in?

CUNNINGHAM: He increased production for his giga-factory up in Reno to produce these advanced lithium batteries. They're great. We like them, but there are impacts. There's huge beautiful basins with playas and the playas have a lot of minerals in them, and to increase the mineral extraction from these playas they're putting in new wells all over and pumping groundwater like crazy. Now they're flooding new areas to try to evaporate lithium out of these salts. Again, there's resource extraction. We're mining lithium to make batteries.

Kevin and I are definitely into solar. We are solar energy advocates and renewable energy advocates but we need to be aware of all the impacts of our lifestyle. Again, it would be really nice to reduce our use. And recycling the batteries: do they just get dumped into a canyon in Mexico or do they really get recycled?

EMMERICH: That's the big deal about these big projects. They have an impact. They have a big carbon footprint. It's almost hypocritical to say the Ivanpah project is off-setting anything.

CUNNINGHAM: They shipped those garage door-sized mirrors—hundreds of thousands of them—from Germany. There's a carbon impact there. We're trying to ask what the carbon offset would be: building this big plant that scraped up the biological soil crust which stores carbon, all these shipping and transportation carbon costs, the metal and cement.

SONNENBLUME: It starts in the red, basically is what you're saying. So how long does it have to be up and running before it's made up for what it cost, carbon-wise?

CUNNINGHAM: Yes. there are people trying to study that, but right now it's a big experiment. So we're saying, slow down bulldozing the desert until we understand what we're losing.

SONNENBLUME: "First, do no harm."

CUNNINGHAM: Right.

May 11, 2016

On August 6, 2019, I spoke with Cunningham and Emmerich on the phone for a quick take on what's changed since Trump took office.

SONNENBLUME: When Trump was elected, one of the very first thoughts I had was, well, this guy's against renewable energy so maybe he'll cancel a lot of the programs that are tearing up the desert, and, if he doesn't slow that stuff down, then we would know that it wasn't about renewable energy in the first place, but was about getting money to corporations.

EMMERICH: That's a very good question and it's complicated, but I can take a stab at it. When Trump got elected, renewable energy on all levels did take a hit. Big projects on public lands were not really approved for a little over a year because they were re-evaluating them. They did this for two reasons, in my opinion. Reason number one is that the administration is very fossil fuel biased. Reason number two was that they wanted to re-write everything that the Obama administration did. Also, the Trump administration proposed a lot of different tariffs on solar panels and that made it difficult for some of the projects to be as affordable as they expected. They threatened to pull some of the tax credits as well.

However, there were ways they helped, too, inadvertently. There's so much opposition to the Trump administration that a lot of states went ahead [with their own programs]. California passed a 100% renewable energy RPS [Renewables Portfolio Standard]. Nevada passed a 50% RPS which actually accelerated applications for big solar with some wind. They did this as part of a backlash against Trump. [An RPS requires utilities to procure a minimum percentage of their energy from sources that are certified by that state as renewable. In the case of California, the goal is to reach 100% by 2045. —KtS]

Some of the public lands projects that are being proposed—and most likely will be built—are taking advantage of the Trump administration's streamlining of environmental reviews.

So the answer is yes and no. The Trump administration did slow it down. But they removed a lot of regulations that were tooled to slow these things down, so it's complicated.

SONNENBLUME: So it sounds to me like some of the direct encouraging —like the funding, the tax help, etc.—was discouraged but that the loosening o f the regulations turned around and encouraged it. The same way that his loosening of regulations is encouraging oil exploration.

EMMERICH: Definitely. Renewable energy projects under the Interior Department are called "infrastructure projects." That's why they get that treatment. Any kind of energy project—oil, gas, renewables—is infrastructure.

CUNNINGHAM: Yes, so Trump isn't giving particular companies giant grants and DOE [Department of Energy] loan guarantees like Obama was, but the price of solar panels has come down so much that they can pretty much build without all the incentives. They still get a tax incentive and Congress has to vote to renew that pretty soon. But Trump is basically saying we want "all of the above" energy including renewable because it's Capitalism. So now we're seeing a whole spate of new projects in the last year. It's not as fast as it was in 2010, but it's definitely becoming a problem for public lands and Desert Tortoise now, even under Trump. All the big "green" groups are supporting bulldozing the Mojave Desert still. They haven't learned anything in ten years. So Basin & Range Watch is again fighting this one by ourselves. It's like business as usual.

SONNENBLUME: When it comes to environmental issues, there's been a real lack of resistance to what Trump's administration is doing.

CUNNINGHAM: Yeah.

EMMERICH: Yes. I'm a little baffled by that. Maybe people are accepting more or think they can't do anything. One theory I have is that it's really going to help a lot of of Democrats—developer Democrats—to roll back a lot of regulation. And when I see some environmental groups getting behind a large build-out of renewable energy here in Nevada, a lot of them don't know the regulations like they should.

CUNNINGHAM: It's troubling, too, because we're seeing Trump rolling back all these energy efficiency policies and regulations and then meanwhile all the national green groups are not, for the most part, lobbying to push for rooftop solar, distributed generation energy storage and energy efficiency. They seem to be laser focused on getting, like, a Gemini solar project on 7000 acres of Desert Tortoise habitat to fight Climate Change. The people we know are not saying, "We need energy efficiency," "We need rooftop solar," "We need to preserve our desert lands which sequester carbon." We have Trump attacking energy efficiency and very few environmental groups trying to rescue that. Like it's too boring or something. To me, it's a little hypocritical. We're absolutely going to need to do that to lower our carbon footprint.

Cross Country Sprint, Part 1

In the summer of 2018, I roadtripped through almost forty of the US American states. I departed from a rural property in southwestern New Mexico that a friend had recently purchased. I'd been helping him set the place up since spring and was planning to go to California next for work but he suggested I delay that. He knew I'd been wanting to take a cross-country trip and he urged me to do it right away. "Things are changing fast," he said. "Maybe you won't be able to do it next year." He was serious enough that he funded it. The amount of cash he handed me was adequate for a few months the way I travel, though for a middle-class professional couple who stays in hotels, eats out, sight-sees and buys gifts, it only would have covered ten days.

I was very grateful for the gift. I knew the trip would affect me in ways I couldn't predict that moment, and would serve as a turning point. That was exactly my friend's intention, and that's exactly how it went down.

TEXAS

August 2nd-4th

Right after crossing the border from the Land of Enchantment into the Lone Star State, I saw this sign: "Welcome to Texas. Drive Friendly—The Texas Way." Shortly after that the speed limit was posted: 80mph. So apparently one can be fast *and* friendly, which was nice to know since nearly all the drivers were exceeding this speed As for my 1986 Toyota pickup, 80 is where it maxes out (unless there's a headwind), so I couldn't be any more friendly than that.

I entered the state near El Paso on the I-10, but switched to the I-20 to take the more northerly route through the state. The landscape was mostly flat for the first few hours, though the horizon was marked by the dark silhouettes of distant mountains, some of them in Mexico. It was hot as hell and the kombucha that my friend gave me as a parting gift was a welcome refresher. In fact I switched to kombucha from coffee as my beverage of choice for long distances on this roadtrip. It gave me a good zing but without the jitters or stomach rot.

I-20 passed through Odessa and Midland, towns I associate with the Bush family and their oil fortunes. What a starkly different place from their New England home! They were following the money, of course, and it worked out for them. I've heard that W's Texas ranch is entirely off-the-grid and runs on renewables, so it's not like they don't know what's up or how to take care of themselves.

I stopped at a gas station and perused a rack of stickers meant for putting on your truck. Besides the inevitable naked lady silhouettes and skull-n-crossbones, there was quite a range, from the innocuous: "Live, Laugh, Love," "Work Hard, Party Harder" and "Wishin' I Was Fishin'"; to the lewd: "I heart Boobs"; to the obstinate: "I'm Only Driving This Way To Piss You Off"; the religious: "God Rules"; the proud: "US Army" and "Iraq War Veteran"; and the racist: "All Trucks Matter."

Then there was the mysterious, "Black Smoke Matters," which I took as racist, but didn't really understand. I posted a photo to social media and a friend explained that "black smoke" refers to diesel exhaust and that some truckers do something called "rolling coal" where they purposely make black smoke pour thickly out their vertical exhaust pipes.

According to the New York *Times*:[1] "Depending on whom you ask, rolling coal is a juvenile prank, a health hazard, a stand against rampant environmentalism, a brazen show of American freedom. Coal rollers' frequent targets: walkers, joggers, cyclists, hybrid and Asian cars and even police officers."

But "Black Smoke Matters" is more than that; it's a Facebook-organized protest effort made up of truckers who oppose a federal law that took effect in late 2017 requiring all trucks to have electronic logging devices (ELDs). ELDs track miles, hours, and breaks as part of an effort to enforce mandatory limits on time worked or distance traveled in one stint. Truckers don't appreciate ELDs because they infringe on their freedom of choice and, if the ELDs register any rule infringements, it is the trucker who is fined, not the company that's pushing the trucker to make unrealistic time. Like so many other workers, truckers have been getting squeezed harder by corporate masters who prioritize profit over people. In my experience, truckers tend to be some of the best drivers on the road in terms of signaling, safety and consistency (with a few obnoxious exceptions), and I'm opposed to making their individual lives more difficult.

All that being said, calling your group "Black Smoke Matters" is tone deaf at best and just plain racist at worst. Neither is acceptable. An anti-environmental sentiment is also obvious and I have no tolerance for that either. So yeah, while there's a trucking industry, let's not make life hell for truckers, but truckers need take some steps in *this* direction, too. Why? Because a) we're all in this together, and, b) there are no jobs, including trucking jobs, on a dead planet.

It took two long days to cross the state of Texas. The first night I checked into a motel just as a dramatic thunderstorm started dropping rain. The lightning illuminated ridges with pointy peaks and flats covered in scrubby vegetation. I was worried about tornadoes but that storm didn't spawn any, at least where I was. On the second day, the landscape turned to rolling hills with oak trees and the green was a welcome change. The drive through Dallas/Fort Worth was less enjoyable. I spent a second night at a motel just off the interstate and crossed into Mississippi the next morning.

THE SOUTH

August 4th-8th

The only way I stayed cool in the South was to drive with all the windows down, all day long, until sunset. After dusk, it was still warm but "sticky" relinquished to "sultry," which was far preferable.

One thing I wasn't expecting was how loud the nights were. Whether I stayed in a campground or a motel, the air was filled with an impressive din when the sun went down. I asked a couple locals what I was hearing but both only said "frogs and bugs"; for them, the sound was just normal background noise and they didn't pay it much mind.

But I was amazed. This, of course, is one of the pleasures of travel: encountering the exotic. Seeing, hearing, smelling, tasting or feeling something new is an enlivening experience, quite apart from whether or not it's fun. Indeed, realistic travelers don't expect a laugh-a-minute. Smart ones know that's not the point. Wise ones treat it all the same.

By placing oneself into a new context, one learns about oneself, naturally. That is to say, for me, every experience is an opportunity not just to do or see something in the world but to explore how I, myself, function as a living creature in different sets of circumstances. I love the great outdoors, but my curiosity runs with equal interest to the deep inside.

As part of a pretentious art-project, I once wrote:

A color is not a constant. Sunlight lends a different hue to your eyes than a General Electric 60 watt or a blinking neon "OTEL" sign. Entering a domed stadium can feel like going inside TV. Rip up a piece of the astroturf and watch the green change when it is on your coffee table. Hold it up next to the screen when you see a game on ESPN. Do they match? Take it with you into the bathroom, throw a towel over the door crack, and shut off the lights. What color is it now? How about when you strike a match? Would it have any color at all under starlight if the moon were new?

It's not just colors that are so mutable to their surroundings, but us. Babies born in Hyderabad and in Atlanta on the same day will lead very different lives. "Nurture" makes an undeniable mark.

Yet nature bestows us with indelible ones, too, but which ones are they? By exposing ourselves to different wavelengths—of light, of culture —we can try to make such distinctions. Various circumstances will throw particular aspects of ourselves into high contrast, while rendering others virtually invisible.

Intention is indispensable to these efforts. Practice also helps. And life springs epiphanies on us from time to time. "Chance favors the well-prepared," it has been said. Yes, and the attentive.

If that makes life sound like a bunch of hard work, I'll say sure, but what else are you doing with your time? "Don't use this life as a waiting room," said the Indian teacher, Osho. Wise words.

So I took great pleasure in the riotous chorus of frogs and insects. In one laid-back National Forest campground in Virginia, I had the joy of hearing them with the sound of running water as I fell asleep. It was such a pleasure that I stayed there three nights. After a week of long driving days and motels, it was also nice to stop moving for a minute.

The campground was dense with vegetative life: trees and shrubs covered in clambering vines and all kinds of flowers grew in dense tangles. These walls of green divided the individual campsites from each other,, providing a pleasant privacy.

This was typical of the South as a whole: so verdant, lush, and *thick*. I tied to imagine how alien this landscape must have been to the waves of European colonizers who arrived between the 16th and 18th Centuries. I could see how some of them considered it to be a kind of paradise.

I also reflected on all the people who lived here before the European invasion. The few who survived disease and slaughter were forced to leave the only home they had ever known and go to places entirely foreign to them, like Oklahoma. The infamous Trail of Tears was a horrific event, but only in degree not kind, and many other crimes of equal or greater cruelty were committed against Native Americans over the centuries. When people say Trump is the worst president ever, my mind first goes to Andrew Jackson; the Supreme Court declared his 1830 policy of Indian Removal to be unconstitutional but he defiantly pushed ahead with it anyway. To discount the brazen offenses of "Old Hickory" is a grave insult to all the Cherokees, Choctaws, Chickasaws, Creeks and Seminoles who suffered so grievously. As bad as Trump is in his own way, he has yet to commit acts this heinous or to flaunt authority this flagrantly. It is certainly telling, though, that he hung a portrait of Jackson in the Oval Office.

The different tribes wildtended the landscapes and farmed their crops in relative harmony with each other and the non-humans. They lacked the gruesome behavior of the empires in Central and South America, were more sedentary than the migratory bands in the West, and were certainly less warlike than their contemporaries in the "Old World." Compared to the Europeans who displaced them—or to us now—their lifestyles were the paragon of "sustainability." It has even suggested that the mound-building tribes of the Mississippi and Ohio river valleys who lived in urban environments dependent on large-scale agriculture (such as the well-known Cahokia complex) voluntarily stepped down when the drawbacks of mass-scale agrarianism became apparent and then returned to simpler ways.[2] Think what we could have learned from these people if we had approached them with open minds and hearts instead of crosses and swords.

* * *

I couldn't spend time in the South without thinking about race. Blacks make up a far higher percentage of the population in those states than anywhere I had lived in the North and I was aware of this whenever I stopped. I enjoyed the feeling of being out of place.

I stayed one night in a motel on the outskirts of Birmingham and I recalled Martin Luther King Jr.'s famous "Letter from Birmingham Jail." This famous piece of writing is exactly what it sounds like: correspondence penned by King while imprisoned. He was arrested on April 12, 1963 during his participation in the Birmingham Campaign.

The Birmingham Campaign had been launched nine days earlier, on April 3, by the Alabama Christian Movement for Human Rights (ACMHR). That day, they released the "Birmingham Manifesto"[3] which began with the words:

> *The patience of an oppressed people cannot endure forever. The Negro citizens of Birmingham for the last several years have hoped in vain for some evidence of good faith resolution of our just grievances. Birmingham is part of the United States and we are bona fide citizens. Yet the history of Birmingham reveals that very little of the democratic process touches the life of the Negro in Birmingham. We*

have been segregated racially, exploited economically, and domi-nated politically.

The Manifesto then lists the various tactics already undertaken, includ-ing petitioning the city, taking their case to the courts, and negotiating with local businesses, but all to no avail. "We hold in our hands now, bro-ken faith and broken promises." Therefore, the time for "direct action" had arrived. "We act today in full Concert with our Hebraic-Christian tradition, the law of morality and the Constitution of our nation," the Manifesto de-clares, and issues an invitation:

We appeal to the citizenry of Birmingham, Negro and white, to join us in this witness for decency, morality, self-respect and human dig-nity. Your individual and corporate support can hasten the day of 'liberty and justice for all.'

The direct action campaign included a boycott of downtown business, marches on City Hall, sit-ins at lunch counters and libraries, kneel-ins at churches, and attempts to register voters at county offices. The city re-sponded by making hundreds of arrests, but the campaign continued. Af-ter a week, on April 10th, the city government of Birmingham obtained an injunction from a state court that allowed it to forbid the protests. Anyone who marched in the city after that point could be arrested.

King's Southern Christian Leadership Conference (SCLC) had been a partner in the Birmingham campaign at the explicit invitation of ACMHR, though King himself had not participated in person up to this point. The city's action forced a decision of whether to continue or not. During plan-ning for the actions, organizers had already agreed not to respect such an injunction since previous campaigns had faltered when they had done so. However, when the moment arrived, there was some hesitation to subject themselves to being jailed; the campaign had run out of money to bail out arrestees, and King was an effective fundraiser who could be useful out-side. After consideration, though, King decided to go to Birmingham any-way: "I don't know what will happen; I don't know where the money will come from. But I have to make a faith act."[4]

So on April 12th—which happened to be Good Friday—King was ar-rested along with fifty others. By prearrangement, organizers refused to of-fer bail for his release and used his imprisonment to gain publicity for the cause. This worked: the story soon went national.

While King was in jail, a group of white clergy in Birmingham pub-lished a condemnation of the direct action campaign in the Birmingham *News*. King received a copy of the paper in his cell and began writing his response in the margins (later finishing it in a notebook provided by his legal team). The letter was completed on April 16th and was published widely.

Like much of what poured forth from King, the letter is eloquent in its language and expresses a clarity of vision that is rooted in both intellectual and moral traditions, while facing the challenges of the present squarely. Everyone who yearns for justice should read the letter in its entirety, as much of it remains relevant to contemporary struggles. What follows are the passages that stood out most for me.

On the need for direct action:

Lamentably, it is an historical fact that privileged groups seldom give up their privileges voluntarily. Individuals may see the moral light and voluntarily give up their unjust posture; but, as Reinhold Niebuhr has reminded us, groups tend to be more immoral than individuals. We know through painful experience that freedom is never voluntarily given by the oppressor; it must be demanded by the oppressed.

Surely King was aware that he was here echoing Frederick Douglass—well-known 19th Century abolitionist and escaped slave—who famously said: "Power concedes nothing without a demand. It never did and it never will." King's distinction between the contrasting likelihoods of individuals and groups to wake up and change their ways stands out and, insofar as it reflects his own experiences and observations in the civil rights movement since the mid 1950's, deserves serious consideration. In our own time, when so much is virtual—including communication and notions of community—are groups even more "immoral"?

On the timing of current actions in Birmingham:

There comes a time when the cup of endurance runs over, and men are no longer willing to be plunged into the abyss of despair.

Haven't we seen this many times since then? The urban uprisings that followed King's assassination were certainly examples of that cup running over. In our own time, an unwillingness to live in despair has sparked street actions against racist policing, especially the ongoing epidemic of cop-on-civilian violence..

On the subject of breaking the law:

One may well ask: "How can you advocate breaking some laws and obeying others?" The answer lies in the fact that there are two types of laws: just and unjust. I would be the first to advocate obeying just laws. One has not only a legal but a moral responsibility to obey just laws. Conversely, one has a moral responsibility to disobey unjust laws. I would agree with St. Augustine that "an unjust law is no law at all."

King defines a just law as one that "squares with the moral law or the law of God" and an unjust one as "out of harmony" with the same. In discussing how laws supporting segregation are unjust he draws on both Jewish philosopher Martin Buber and Christian theologian Paul Tillich. His level of discourse on this point—and on others throughout the letter—displays his academic background, which included a doctorate. This is not writing you can just skim. Nor could it be reduced to a series of tweets.

Bringing it back to the specific case of his arrest:

> *Sometimes a law is just on its face and unjust in its application. For instance, I have been arrested on a charge of parading without a permit. Now, there is nothing wrong in having an ordinance which requires a permit for a parade. But such an ordinance becomes unjust when it is used to maintain segregation and to deny citizens the First-Amendment privilege of peaceful assembly and protest.*

Debates over whether to obtain permits for political marches continue to this day. Certain types of organizers still fetishize legality, and often they are the same ones who profess non-violence in the names of Gandhi and King; I suggest it would be useful for them to revisit King's views on the subject. Rather then being opposed to breaking the law *on* principle, he is in favor of breaking it *as* a principle:

> *One who breaks an unjust law must do so openly, lovingly, and with a willingness to accept the penalty. I submit that an individual who breaks a law that conscience tells him is unjust, and who willingly accepts the penalty of imprisonment in order to arouse the conscience of the community over its injustice, is in reality expressing the highest respect for law...*

Here is arguably the heart of the philosophy of non-violent civil disobedience: the appeal to conscience. Notice that King speaks here of the conscience "of the community," not of the oppressor, the oppressor's class or the oppressor's power structure. Critics of non-violent tactics—such as Craig Rosebraugh, in his book, *The Logic of Political Violence*—have rightly pointed out that one's oppressor might lack a conscience to appeal to. King apparently understood this, advocating instead to go over the head of the government to the people.

King's critics in the Birmingham clergy were white, and he takes direct aim at them and their class here. Here are the kinds of words and sentiments that are still regularly skipped over in favor of snippets from the "I have a dream" speech:

> *I have been gravely disappointed with the white moderate. I have almost reached the regrettable conclusion that the Negro's great stumbling block in his stride toward freedom is not the White Citizen's*

Counciler or the Ku Klux Klanner, but the white moderate, who is more devoted to "order" than to justice; who prefers a negative peace which is the absence of tension to a positive peace which is the presence of justice; who constantly says: "I agree with you in the goal you seek, but I cannot agree with your methods of direct action"; who paternalistically believes he can set the timetable for another man's freedom; who lives by a mythical concept of time and who constantly advises the Negro to wait for a "more convenient season." Shallow understanding from people of good will is more frustrating than absolute misunderstanding from people of ill will. Lukewarm acceptance is much more bewildering than outright rejection.

Those are some hard-hitting words that have lost none of their relevancy—or punch—in our current day. If anything, the situation is worse now, with the Obama presidency having served as a rationalization for many white moderates to dismiss the relevancy of racism as the persistent problem it is. Today's white moderates took the sight of black grandmothers crying with joy on the night of the 2008 election as proof that "the problem of race" had been solved. What more could they want? Surely this showed that we could all move on now, right?

Black Lives Matter put the lie to that narrative. They knew "shallow understanding" and "lukewarm acceptance" when they saw it. The fact that the most vocal Black organizing in a generation erupted when the White House was occupied by its first Black family was itself a clear, unequivocal message. As King wrote: "Oppressed people cannot remain oppressed forever. The yearning for freedom eventually manifests itself."

King understood that the "yearning for freedom" in the United States was connected with movements abroad:

Consciously or unconsciously, he has been caught up by the Zeitgeist, and with his black brothers of Africa and his brown and yellow brothers of Asia, South America and the Caribbean, the United States Negro is moving with a sense of great urgency toward the promised land of racial justice.

This international perspective distinguishes King's analysis from that of most activists in the contemporary US, who tend to be blind to foreign policy and limit their focus exclusively to domestic issues. Witness how the emphasis on wages, health care, debt, and even Climate Change ignores the issue of US imperialism. King would not take up that topic explicitly and publicly for another four years (in his "Three Evils" speech at Ebenezer Baptist Church, a year to the day before his assassination), but here we see that his perspective was already more expansive than the borders of the US.

People today might find it hard to believe that King was ever called an "extremist" (especially given the watered-down version of his beliefs that has become prevalent) but he was, and he addresses the term directly:

So the question is not whether we will be extremists, but what kind of extremists we will be. Will we be extremists for hate or for love? Will we be extremists for the preservation of injustice or for the extension of justice? ...Perhaps the South, the nation and the world are in dire need of creative extremists.

"Creative extremists"—there's a label worth taking up again! Given the multiple crises of our time, which are themselves extreme, that's what we should be aspiring to be.

King speaks about the limited outlook of those on top of the social hierarchy, and this too has not changed:

Few members of the oppressor race can understand the deep groans and passionate yearnings of the oppressed race, and still fewer have the vision to see that injustice must be rooted out by strong, persistent and determined action.

King then devotes a long section to criticizing the churches that had so far refused to stand with him and with everyone else fighting for civil rights. Here are two pointed excerpts:

I felt that the white ministers, priests and rabbis of the South would be among our strongest allies. Instead, some have been outright opponents, refusing to understand the freedom movement and misrepresenting its leaders; all too many others have been more cautious than courageous and have remained silent behind the anesthetizing security of stained glass windows. ...In the midst of blatant injustices inflicted upon the Negro, I have watched white churchmen stand on the sideline and mouth pious irrelevancies and sanctimonious trivialities.

In King's mind, the church had a historical role as a rabble-rousing institution, not as a supporter of the status quo, and that by abandoning that role, it was putting its own existence at stake:

If today's church does not recapture the sacrificial spirit of the early church, it will lose its authenticity, forfeit the loyalty of millions, and be dismissed as an irrelevant social club with no meaning for the twentieth century. Every day I meet young people whose disappointment with the church has turned into outright disgust.

Enigmatically, he then adds:

...Perhaps I have once again been too optimistic. Is organized religion too inextricably bound to the status quo to save our nation and the world? Perhaps I must turn my faith to the inner spiritual church,

the church within the church, as the true ekklesia *and the hope of the world.*

"Ekklesia" is a Greek word—*εκκλεσια*—that appears in the New Testament of the Christian Bible usually translated as, "church," and which is a compound word made up of "ek-," for "out" and "kalein" for "to call"; that is, to "call out" or to "summon." The early Christians borrowed the term from the ancient Greeks, for whom it designated a public assembly of citizens convened to deliberate political issues.[5] "The church within the church" is an intriguing phrase, and King did not coin it. The first "church" describes a spiritual community and the second an institution; that is, the people who live as believers as opposed to the structures that dictate dogma. King is implying that, if need be, he is willing to walk away from establishment Christianity in favor of the Christianity that exists among believers, unorganized. For some people in his audience, this might have been the most controversial statement he made in the entire letter.

King was devoutly Christian which gives his words a particular punch. Earlier in the letter, when he criticized the churches, he was clear: "I say this as a minister of the gospel, who loves the church; who was nurtured in its bosom; who has been sustained by its spiritual blessings and who will remain true to it as long as the cord of life shall lengthen." I am not myself a believer, and I am well aware of all the great evil perpetrated in the world in the name of Christianity. Gore Vidal was not far off the mark when he declared that "the great unmentionable evil at the center of our culture is monotheism."[6] Yet when I read King's words, or hear one of his speeches, I cannot lump him in with the brutal Spanish conquistadors of the European invasion, the bloodthirsty Crusaders of the Middle Ages or the bigoted fundamentalists of our own time. And when it came to that last group anyway, King made it abundantly clear that neither could he. For whatever reason—his education, his upbringing, his inborn constitution—he was able to draw on Christianity to fuel his fight for justice. I imagine that if he had he been born into a different time and place cursed with oppression but without Christianity, surely he would have drawn his inspiration elsewhere.

Putting the local struggle into a wider context, both in place and in time, King wrote:

> *I have no fear about the outcome of our struggle in Birmingham, even if our motives are at present misunderstood. We will reach the goal of freedom in Birmingham and all over the nation, because the goal of America is freedom. Abused and scorned though we may be, our destiny is tied up with America's destiny.*

King wrote these words over half a century ago, and at this point in time, having one's destiny tied to America's seems dangerous. The nation is on a headlong rush toward disaster, politically, economically and ecologically, and is threatening to take down a bunch of other lives with it. It strikes me that the prudent course would be to untie oneself from "America" at this point, if possible.

I didn't know what to think about King's phrase, "the goal of America is freedom," so I turned to activist Forrest Palmer for his opinion. He replied:

> That is a lie. I don't know if he was saying that because [he] actually believed it or if it was something to ingratiate the white power structure that controlled his personal and professional life as well as those of his people. Whatever the reason for such a misguided statement, it was still misguided.
>
> I think that people need to start analyzing things in an honest retrospective capacity and not one that is totally beholden to individuals we consider hero figures. It is not a dishonor to go back and critique people when we have more information at hand.... We can't just relegate our present circumstances to the positions of the ones who preceded us since the times, they are a changing.

Finally, King had a few words about the police, and those who approved of their behavior:

> You warmly commended the Birmingham police force for keeping "order" and "preventing violence." I doubt that you would have so warmly commended the police force if you had seen its dogs sinking their teeth into unarmed, nonviolent Negroes. I doubt that you would so quickly commend the policemen if you were to observe their ugly and inhumane treatment of Negroes here in the city jail; if you were to watch them push and curse old Negro women and young Negro girls; if you were to see them slap and kick old Negro men and young boys; if you were to observe them, as they did on two occasions, refuse to give us food because we wanted to sing our grace together. I cannot join you in your praise of the Birmingham police department.

In the present day, police are often complimented by the media or by liberal organizers for their "restraint," even in situations when they showed little to none. This is exactly what King was talking about. He also says that, when the police *did* act "nonviolently" it was for one purpose: "To preserve the evil system of segregation," which is, of the status quo. He goes on: "I have tried to make clear that it is wrong to use immoral means to attain moral ends. But now I must affirm that it is just as wrong, or perhaps even more so, to use moral means to preserve immoral ends."

With a characteristically poetic flourish, King signs off:

Let us all hope that the dark clouds of racial prejudice will soon pass away and the deep fog of misunderstanding will be lifted from our fear drenched communities, and in some not too distant tomorrow the radiant stars of love and brotherhood will shine over our great nation with all their scintillating beauty.

Amen to that.

Manhattan

August 11th

Manhattan is overwhelming to the senses at two scales simultaneously. First, it is monumental—in the sprawl of its blocks and the heights of its structures. I felt like such a tiny creature among the giants of glass and stone. Ten stories is enough to dwarf a person, and a whole block of buildings even half that high on both sides makes a canyon. Many buildings in NYC are much, much taller, of course.

Secondly, it is overwhelming at the close-up, human scale. Every surface has been touched an uncountable number of times. Maybe you are the 50,000th person to sit in a particular cab, and how many hands have pressed that "walk" button? Literally millions. How many feet have touched the sidewalks? *Hundreds* of millions. Every wall at eye-level has been *used* over and over: leaned on, pasted over, taped, kicked, peed on, painted, spat on, scrubbed (maybe) or otherwise messed with, *ad infinitum*. Words and symbols clutter the field of vision: signs on buildings, vehicles and posts, screens with moving images. More than you could ever read or take in without stopping, and stopping often and for awhile. Buy this, buy that, stop, go, come in, stay away. It's a din to the eyes.

Then there are all the windows. The word "window" is Norse in origin, and was originally, "wind eye": the eye of a house, through which the wind blows. Of course, on many buildings constructed in the last half century, the windows don't even open anymore, but nonetheless, they remain eyes, eyes that go both ways. From outside they show what's inside; the storefront display is the best example. On upper floors you see evidence of domestic life: lamps, plants, cats. Here and there a pane is open and through it a curtain blows or music wafts (or pounds).

One can view the history of architecture in Europe and the USA as a progression of enlarging window size. Before the age of electricity and central air, windows were necessary for lighting and ventilation, and couldn't be skipped. With masonry construction, such as stone and brick, this was challenging since the walls supported all the weight of everything above, including the roof, so the taller the building, the thicker the walls needed to be at ground level, and any gaps were points of weakness. The amount of load-bearing that could be spared for openings was definitely finite. Think of old churches, like cathedrals, and how narrow their windows are. They had to be. Most of the walls' bulk was needed to hold up the roof at its desired loftiness. Even with limited openings, the height of the walls still proved too much and flying buttresses were invented to support them.

The big game changer was the development of steel-frame construction in the late 1800's. According to legend, the inspiration came when an engineer who was fretting over how to make buildings taller came home from work one night and set a stack of heavy books on top of a bird cage by the door. Though the structure was made up only of narrow metal bars, the weight of the books was evenly and—this is key—*safely* supported. Replace the bars of the cage with steel beams, and you have a watershed moment. Even if this story is apocryphal, it's an excellent illustration of the concept, so we'll go with it.

The strength of this technique is provided entirely by its metal "skeleton" and so the "skin" could be anything, from granite to terra cotta to glass. Three quarters of a century passed before windows could be produced at the size necessary to clad a building with nothing else, but once they were, the "glass box" office building became the standard for several decades. This style was called "Modern" in the United States, and was characterized by a lack of decorative ornament; a style of no style, in a way.

NYC, like most US cities, has a plethora of glass boxes. Some are sheathed in privacy by highly reflective sheets that change with the weather and, depending on your taste, can be beautiful when struck with certain qualities of light. Personally, I am amazed at such structures, but don't find them very attractive. *Inside* these buildings is where you want to be—if you have to be inside a building at all—because all the windows are a welcome feature then. If your view is of older buildings with cornices, arches, pilasters, bow fronts and rusticated foundations—*i.e.,* ornament— so much the better.

One can wonder how medieval builders would redesign cathedrals if steel-frame technology were shipped back in time to them. Imagine Notre Dame with glass in the place of stone. Not that it's as simple as swapping one material for another, though. With a metal skeleton holding up the ceiling, the famous Gothic arches would be totally unnecessary, as would the soaring curves of the flying buttresses. Even the cross-shaped layout could be discarded. Certainly, famous American architect Louis Sullivan was correct to suggest that form should follow function. However, form has no choice but to follow material, and, freed of the constraints of masonry, form can take on nearly any shape (as contemporary architect Frank Gehry has vividly demonstrated with his bold designs, in which regular shapes have been stretched, twisted and turned halfway inside out).

I was once a great lover of architecture. When I was in grade school, I wanted to be an architect when I grew up. By high school, the love had become strong enough that I was incensed when Omaha, the city I lived in, tore down large section of historical buildings downtown to make way for a "campus style" corporate HQ for the Con-Agra corporation, with lots of grassy berms and short structures. A friend and I spent the better part of a day shortly before demolition walking around the district, which was known as "Jobber's Canyon" and which was the largest extant example of what is called an "urban canyon." I snagged several big elevator gears as souvenirs and they ended up as yard art at my parents house quietly rusting among landscaping plants.

The city later regretted its choice. In the 2000's, fashions changed and people started returning to downtown cores in droves all over the country. Jobber's Canyon was exactly the type of neighborhood that attracted this demographic and their investment. Con-Agra had moved out at some point so the whole "development" was especially pointless.

My love of what urban planners call "the built environment" played a key role in the formation of my critiques of US culture. When I moved to Minneapolis in 1993, I adored the old architecture there, and was delighted to move into a 1920's-era apartment building with hardwood floors, built-in glass-doored cabinets, and a fireplace. Whole neighborhoods of such places existed around the downtown.

But it didn't take me too long to note the damage done to the city by the urban renewal projects and highway building of the 1950's and 1960's. Close to my home was the interchange of I-35 and I-94. Each one approached the spot in huge trenches more than a block wide. At this intersection, a huge area of the city had been cleared and excavated for a tangle of ramps and overpasses. You could only have done more damage to the former neighborhood by bombing it. I also noted that the path of I-94 west of this location followed along the line where the the old grid of Minneapolis—parallel with the river, and so at about a 30° angle from true north—met the new grid, which was set exactly to the compass. The district that had been wiped out had been made up with one intersection after another that wasn't a simple 90° meeting, each one instead made up of all sorts of acute and obtuse angles. How funky! There must have been some creative architectural responses to these oddly-shaped lots, but now it was gone, and even if you capped the highway and knitted the fabric of the city back together there, you could never replace what was lost.

But it got worse. East of the junction, I-94 had torn right through the heart of "Frogtown," a predominantly African-American district. A thriving business district was destroyed and a neighborhood hopelessly disrupted, all to make it easier for white suburban commuters to get to their jobs downtown.

So I looked at the beauty that had been heartlessly destroyed for reasons of greed and politics and I saw how the US system worked: it was callous, short-sighted, and racist.

At about this time, James Howard Kunstler released his book, *The Geography of Nowhere*, in which he spelled out exactly why the suburbs, modern "development" and the destruction of historical buildings sucked. So it wasn't just a matter of taste. It was a fact that city life in the post-WWII era had taken several turns for the worse, the relative wealth of the time (for those not living in red-lined neighborhoods anyway) not withstanding.

When I moved to Boston in 1995, I was blown away by a place that had suffered grievously from urban renewal and highways but remained a showcase of every architectural style that had existed in the US since the late 1600's. Taking a walk there was like flipping through a textbook of historical periods, and much remained delightfully intact, including the Victorian-era Back Bay, the ritzy Beacon Hill and my own neighborhood, the North End, which was the oldest continuously-inhabited urban neighborhood in the United States. I lived two doors down from the Old North Church ("One if by land, two if by sea"). My favorite breakfast place around the corner was next to Paul Revere's house. The cemetery in the middle of the neighborhood had gravestones dated to the 17th Century, pockmarked by Redcoats using them for target practice during the British occupation of the city during the Revolutionary War. From that spot, you could look across Boston Harbor to Charlestown, and see the sailing ship, the USS *Constitution*, aka "Old Ironsides," anchored at the docks there.

But after moving out West in 2001 and becoming personally acquainted with monumental landscapes in nature and the intimate intricacies of ecosystems, my attraction to architecture and my fondness for urban areas faded. The awe I once felt in an old church paled to the reverence I experienced in a stand of old growth trees. Likewise, my interest in people, their scenes and urban spaces was dulled by my fascination with wildflowers, the insects who pollinate them, and the ecology that draws them all forth. So this trip to NYC was not as exciting as it would have been twenty years previously.

40

One piece of architecture did cause me to stop and stare in wonder and that was the Flatiron Building. More than the other arts, architecture is wedded to practical concerns, and in this case economics demanded that the use of every square foot of the lot be maximized. The challenge here was presented by the shape of the lot: a triangle formed by the meeting of Broadway and Fifth Avenue at an acute angle of about 25°. The resulting building, designed by famed Chicago architect Daniel Burnham and finished in 1903, was a striking achievement: a 21-story skyscraper that is 87 feet wide at its south end but only 6 feet at its north corner. Originally called the "Fuller Building" to honor the founder of the contracting company that built it, it quickly earned the nickname "Flatiron" for its shape, and this moniker became official in the 1920s.

The Flatiron Building is of steel-frame construction clad in limestone and terra cotta. Burnham employed classically-inspired decorative motifs and proportions as was popular with the "Chicago school" of design at the time. The structure is divided into three parts like an ancient column: base, shaft and capital. Here the base and capital are each four-stories, with thirteen stories of shaft between them. Viewed "head-on" from East 23rd, the building resembles the prow of a ship. The windows on this distinctive corner have curved glass panes.

I stopped and snapped a selfie with the building. In the past, I would have taken a bunch of photos from multiple angles, far away and up close, and seen if I could get inside. But this time it was enough to admire it for a moment and move on.

Not wanting to deal with NYC traffic, I had parked my truck in Newark and taken the PATH train into the city. I disembarked at the World Trade Center and was immediately blown away by the intense stimulation that is NYC. Checking the time, I was a worried about being tardy for a lunch date so I grabbed a taxi. Talking to a New York cabbie also felt like something I should do as a writer.

I was in luck. My cabbie had been driving in Manhattan for over twenty years, so he had plenty of observations and opinions about the city and how it had changed in that time. For one thing, rent had skyrocketed and driven many people out. He himself could no longer afford to live on the island and was now in Queens. Traffic had worsened too, and he hailed the wisdom of leaving my truck in New Jersey. "People are mean to out-of-towners here," he said ruefully. "They should be welcoming and understanding but they're not. They see an out-of-state license plate and they honk, they yell." He had taken us over to the east side highway and we made good time north but once on 23rd Street, traffic was stop and go.

We watched each light change several times before passing through. Before the fare hit $20, I disembarked to walk the rest of the way.

There were plenty of people on the sidewalk, but it wasn't over-crowded and I had no trouble moving at the quick clip I wanted. At intersections, I went with the flow, crossing when there was an opening in traffic regardless of what color the light was. It felt good being back in that zone. Dense, frenetic Boston had trained me well in these ways.

After lunch, I chose to walk back to the WTC from there since it was only about two and a half miles and I wasn't in a rush. I ended up going through Washington Square Park and SoHo. Much of that route was among shorter buildings, and felt homey. I was amused to spot Kale growing in a garden box under a sidewalk tree.

I saw lots of trendy people in hip places and could almost smell the money. Lower Manhattan is no longer a quarter of starving artists, writers and musicians and hasn't been for quite some time now. Once I dreamed of living there but the world I envisioned no longer exists.

VERMONT

August 18th-22nd

Driving through Vermont countryside is like passing by one postcard after another. It's all red barns, white farmhouses, tidy pastures, placid cows and wildflowers. The idealized picture of a rural landscape in the US is based on this typical New England countryside and is presented to us on TV, greeting cards, in movies and on product labels.

One of the strangest sights I've ever seen, and definitely the opposite of ideal, was in the town of Rutland where I stopped to find a coffee shop. I spent some time walking around the downtown, which was typical for the area: charming brick or stone commercial buildings two to four stories in height, with stores on the first floor and apartments or offices above. Nice for what it is (more on that in a minute), and kept up better than in many other places. As I walked down the block, I saw an opening on the next corner, and assumed it was a big square or a park. But when I reached that intersection I stopped in my tracks, shocked at what I saw. Before me was a huge, suburban style parking lot and, set way back from the street, was something only slightly less out-of-place than a UFO would've been: a Walmart.

Walmarts in small towns are hardly unusual of course but they're always on the outskirts, far from the old "main street" core, with a section of post-WWII sprawl between.

But here the big box was literally *on* their main street (here named Merchants Row). It was quite jarring. I wondered how such a thing had ever been allowed. Later research uncovered that the site had formerly been the home of the Rutland Railroad train station, finished in 1854 and torn down 101 years later. Sometime after 1963, a shopping center was built on part of the property and the Walmart went in sometime after 2005 when town residents decided to open the rest of the lot to development.[1] So the marring of their home in this way was a recent event.

I don't want to make "small town America" into something it's not or never was. All such places were founded to facilitate resource-extraction whether that was trapping, farming, mining, logging, fishing, etc., and as such they are the centers of domination for their surrounding ecosystems. All of them displaced habitat with their platting in ways that the migratory camps of gatherer/hunters never did or could.

Yet for all that their scale was still human-sized and walkable. People lived in such places before electrification, in times when the majority of their food came from no more than a day or two away, and when cars and 2500 mile supply chains didn't exist. As such, they represent a better way of living than what we currently practice, though far from the best.

We must also never forget these towns were built on land stolen from one set of people, and the wealth to construct them was generated in part by enslaving of another set of people. This history of white supremacy is inescapable, as is the fact that the current population continues to benefit from its brutality. Reparation has yet to be offered and is still owed. Vermont's population today is about 90% White, with 2% Black and less than 1/2% Native American. So this, too, is part of the cultural "ideal" that is set by the state and its scenery: whiteness.

June 2019

NOTES:

[1] Tabuchi , Hiroko. "'Rolling Coal' in Diesel Trucks, to Rebel and Provoke" (New York *Times*, Sept. 4, 2016)
[2] According to Prof. Randy Woodley as quoted in my "Failures of Farming & Necessity of Wildtending" (Macska Moksha Press, 2018), p. 17.
[3] BHAMWiki. "Birmingham Manifesto."
     http://www.bhamwiki.com/w/Birmingham_Manifesto
[4] King Encyclopedia. "Birmingham Campaign."
     https://kinginstitute.stanford.edu/encyclopedia/birmingham-campaign
[5] Thayer's Greek Lexicon, as quoted by BibleHub. "1577. ekklésia."
     https://biblehub.com/greek/1577.htm
[6] Vidal, Gore. "(The Great Unmentionable) Monotheism and its Discontents" (The Lowell Lecture, Harvard University, April 20, 1992).

The Myth of US "greatness"—
A conversation with
Margaret Kimberley

In mid-August, I enjoyed the great pleasure of meeting Margaret Kimberley for lunch in midtown Manhattan. Ms. Kimberley is an Editor and Senior Columnist for Black Agenda Report, *a member of Administration Committee for United National Antiwar Coalition, and a member of the coordinating committee of the Black Alliance for Peace. She is a contributor to the anthology,* In Defense of Julian Assange *(OR Books, Oct. 2019) and the author of* Prejudential: Black America and the Presidents, *to be published by Steerforth Press in February, 2020.*

I have been an admirer of Ms. Kimberley and her writing for a number of years so it was a thrill to sit across a table from her. Our discussion ranged over many topics, including politics and media, technology and social change, and history and the future. What follows is a transcript edited for length and clarity.

TRUMP, RACISM AND IMMIGRATION

KOLLIBRI TERRE SONNENBLUME: How do you react when people are like—as some political scientists recently said—that Trump is the, quote, "worst president ever."

MARGARET KIMBERLEY: It means they don't like him. That's all it means. "Worst" based on what? Did he invade another country? I mean, look at what they've done. How they've rehabilitated George Bush. Depending on which number you believe, he killed a minimum of half a million people in Iraq. Maybe one million. But Trump is worse? Now maybe before he leaves office he'll do something equally as horrible, but he's certainly not worst right now.

SONNENBLUME: Not yet. It seems to me that it shows very little—

KIMBERLEY: It shows you they're biased. They don't even dislike him for the reasons they should dislike him. Trump was dangerous to them and they were spying on him during the campaign because he talked about ending the neoliberal consensus. Now, he's not a peace candidate, but for him to even say we should have a better relationship with Russia, or, "I wouldn't have gone into Syria," or "These trade deals are bad"—he's up-ending everything that they depend on. It doesn't make him good, but it makes him dangerous to them, and that's why they don't like him. They don't even dislike him because he's a racist.

He is a problem for black people. It is bad when overt racism is given any cover or given any credibility. That is always bad for us. I know people who say they would rather have than an overt racist than a covert racist because you know where they're coming from, but I don't agree with that. I think it's always dangerous when they feel safe. Because then they're emboldened. That's why you had Charlottesville last year.

So that's very dangerous to us. But having said that, all the Democrats offer is "we're not like that." It's like, we're just not overtly racist crackers, and I ask if that's it? That's all you have to stand on?

SONNENBLUME: What do you think about the idea—trying to find a hopeful scenario or whatever—that demographically the United States is changing, will not be a majority white country within the foreseeable future—within a couple decades—and that what we're seeing here is the last hurrah of a group of people who know they're losing?

KIMBERLEY: I think that's one of the reasons Trump got elected. That's why there's so much antipathy against immigrants. I guess you saw the news about his wife's parents getting their citizenship.

SONNENBLUME: [laughs] I did.

KIMBERLEY: Isn't that "chain migration"? She sponsored her parents. It's not about "immigration"; it's immigration of everybody with brown skin. That's what that's about. So yeah, they're terrified. It's going to be harder for us. Things will get worse.

POVERTY AND PROPAGANDA

KIMBERLEY: Very few people will talk about how poor Americans are. But everybody insists on talking about the "middle class"—this mythical term that's meaningless. Americans are poor. I mean, it's that simple. There's job growth but it's low wage work. When I was a kid, and I guess for most of my life, the number one employer was General Motors. Living wage work. Union work. Pensions, benefits. Now the number one employer is Walmart. What does that tell you?

SONNENBLUME: Who tell their employees how to apply for government benefits because they're not paying them enough to afford food.

KIMBERLEY: I think as things worsen, I'm afraid that we will see more acting out; the mass shootings, etc. I'm not really surprised. There are a lot of people who are becoming unhinged because they can't count on the things they could count on. And it takes very little to push them over...

Glen [Ford] wrote so profoundly after the mass shooting in Las Vegas that in this country, people don't have solidarity. Everybody's an enemy... It's difficult to have solidarity in a settler-colonial state because everybody is your enemy.

So I think we will see retrogressing. That's why you have "stand your ground" laws and people wanting to have every kind of weapon, as there's a lot of fear. And they have no other way to cope. They haven't been told that there's any other way to cope. And the big problem for Americans is we're told we have the best of all possible worlds. There's no alternative. There's nothing else to consider. No other country—I mean, just go to Canada. Just go north. They give everybody health care. But people are told we have the best health care system in the world when we don't. Life expectancy is like 30th or something in the world. [The WHO lists the USA at 31 and the UN at 44.]

So here you have this "best," "wonderful," "only" system but it's failing. So then what do you do?

SONNENBLUME: Because it's based on competition, not cooperation.

KIMBERLEY: Competition and white supremacy. So there you are, a white person with expectations about what you are going to get in life, and then you don't get them. But you don't band together with other people to change things. Everyone else is just your enemy. So I think we're going to see, unfortunately, more very bad, very dangerous behavior, I'm sad to say.

SONNENBLUME: ...One disappointment for me is that I was hoping that when Trump got elected, perhaps, there would be a resurgence in antiwar thought because people could be like, "Oh, look what he's doing, so let's oppose that." Great! I was even thinking, "Well okay, maybe I'm even willing to keep my mouth shut about you opposing war now that you said nothing about while Obama was just waging it for eight years, but okay, great, at least you're opposed to it now." But I feel like I see none of that at all. There's been no resurgence in antiwar. What's going on? Why doesn't anyone care?

KIMBERLEY: I think there was a large groundswell of antiwar activism around Iraq because Bush was the president. First of all, because he was a Republican, not a Democrat. And because he was sending US troops. There aren't many people who are against US intervention. So if you can destroy a country like Libya without sending any Americans in combat roles, and you pay some jihadists to do it—which is exactly what they did—Americans don't care that much. So if you're not sending Americans somewhere to potentially get killed, nobody cares. So use jihadists as proxies. Or drone warfare. Or sanctions, which are war by other means. It's easier to cover up and people don't care as much...

Americans are among the most propagandized people on the planet, and they would be terribly offended if you said that to them.

SONNENBLUME: [laughs] They are when I do, yeah!

KIMBERLEY: ...If everything you know about Russia is from the New York Times and the Washington Post, you don't know anything about Russia. Just for example: Ukraine. The president of Ukraine, Victor Yanukovych, was elected. The United States had a coup against an elected president to put in place a bunch of neo-Nazis. But Putin's the villain? Trump is right: Crimea is more Russian than Ukrainian. So when he said, "Well, they mostly speak Russian," he actually, for once, said something that's true. And it's just dismissed. "Oh, he's just listening to Putin." No. Look it up. Crimea is mostly Russian so it's plausible they would vote to be part of Russia instead of Ukraine.

SONNENBLUME: It was 80% or more.

KIMBERLEY: I'm sure it was a majority. Why, as a Russian-speaking person, would you want to remain a part of a Ukraine which banned the teaching of Russian, among other things? You wouldn't. But nobody knows that.

SONNENBLUME: The facts of the matter here are irrelevant. I can't remember who said it, but someone said politics is not about facts, politics is about emotions. So that's what we're seeing here. This demonization of this individual—who, obviously, I can't say enough bad things about him: Trump—but this demonization is completely at the expense of looking at the system. Not looking at anything systemic whatsoever.

Correct me if I'm wrong, but my impression—last year, as we went up to the election, I mean two years ago now—was that I'd never seen the media actually line up behind a candidate the way that they lined up behind Hillary. It really seems as though a large percentage of the media—I mean, certainly the important ones: the Times, the Post, CNN, etc.—were just overtly on her side when usually they at least pretend not to be on one side or the other. I feel like I hadn't seen that before in my lifetime.

KIMBERLEY: Actually, I've seen it a couple of times. The media supported Clinton over Bush, Sr. They were kind of divided with Gore and Bush. Actually there was a lot of Gore antipathy, I felt. They favored Obama over Mc-Cain. It's funny, people always like to say—Obama fans—he didn't have any scandals; there were no scandals. If a president has no scandals, that means the press is covering up for him. So if they were really doing their jobs, there would've been some scandals to report.

THE DECLINE OF THE DEMOCRATS

SONNENBLUME: So thirty years ago, in 1988, was the first election I was old enough to vote in. I was in college at the time in Minnesota and they have caucuses there instead of primaries for the Democratic Party, so I went and caucused for Jesse Jackson. Obviously he didn't get the nomination. But recently I went and looked up his platform and I was astounded at the things that he was running on. He was mentioning reparations [for slavery]! I mean, incredible! I was just wondering if you wanted to say anything about how far they've fallen in thirty years. Sanders was nothing like Jesse Jackson. Nowhere close.

KIMBERLEY: No! It's so funny, the Democrats have moved so far to the right. This idea that Sanders was a Socialist—no, he was an old school Democrat. All Democrats sounded like him forty years ago. That's not Socialism. He's just a Democrat with a little bit of reformism. But that still makes him anathema. So they scapegoated him: Oh, if he hadn't run against Hillary! So you're absolutely right: Democrats now,—supposedly left-wing Democrats, the few who even claim to be on the left of the Democratic Party—now they're like the moderate Republicans of old. Seriously. They're like the Republicans who were in favor of abortion rights. That's it.

SONNENBLUME: Jackson was calling for a cut in military spending and I believe that Mondale was still calling for such a cut. Maybe Dukakis, but it disappeared after that.

KIMBERLEY: Remember Dukakis riding around in the tank. [laughs] You're right. And Bernie, he's not antiwar. He'll say something mild like, "Palestinians deserve respect" and because no one mentions them at all or [only] speaks of them in a racist, horrible way, they say, "Oh, he said he supports the Palestinians." No, he didn't. He's not really anti-war. He only said about drones, "Oh, we did it too often," or something like that. He's fully on board with the whole Russiagate thing. He's pretty much the imperialist. So the only reason to be for him, it seems to me, is because of some improvement on some domestic issues. But he's a far cry from what used to be the left wing of the Democratic Party.

SONNENBLUME: It's not as if anything that Bernie was talking about—free education or expanded Medicare—could happen without cutting the military budget. I mean, that's what has to happen.

KIMBERLEY: Absolutely. You can't have both.

SONNENBLUME: You can't have anything until you cut the military budget.

KIMBERLEY: These Democrats—like the members of the Congressional Black Caucus—they will talk out of both sides of their mouths. On the one hand, they'll say don't cut social program spending, but then they vote for the increase in the military budget, when you can't have both.

And this is the other thing that pisses me off. They always talk about: Trump is horrible, Trump is vile, Trump is the apocalypse, he's the Antichrist. He says let's increase the military budget 10% and most Democrats go along with him. So they don't really hate him. If he had better manners, they would just go along with him. When he did this minimal strike on Syria, that's when Democrats praised him.

SONNENBLUME: I remember that. I was in southern California when that strike happened and I picked up a copy of the L.A. Times and they had a banner headline about it and I can't remember the wording of it, but it was like a headline that would appear on the Sports page. That's the tone they were using describing these strikes against Syria. Since Shock-and-Awe, it's sort of been a video game for people to watch on TV.

KIMBERLEY: Yes. There aren't any Americans over there, so they demonize Assad and tell you he's a murderer and a killer and don't tell you that it's American-backed jihadists who are chopping off people's heads, who destroyed this country. There wouldn't be any Syrian refugees if the United States hadn't decided to pick off Assad. And that goes back to the neo-con Project for a New American Century. Get rid of the secular Arab states. So they get rid of Saddam Hussein, then Gaddafi, and Assad was supposed to be next.

And Russia, they weren't always backing Assad. Putin was ready to shove Assad under a bus, but Obama wouldn't take yes for an answer. They were willing to let Syria to be carved up. They were saying, "Assad has to be the president, but we can talk, we can negotiate something." And then Ukraine happened and they realized, "Well, we can't trust the Americans at all" and that's when they went in.

Russia and China can be blamed for Libya. As permanent members of the Security Council, they could have vetoed the No Fly Zone resolution. But they let it happen. So they cut some sort of a deal: you don't vote against me, I won't vote against you. God knows what the hell they did. And then they realized that they had messed up. So they haven't gone along in the Security Council when it comes to Syria because they realized they messed up in Libya. They see what a disaster that was for their countries.

Remember Gaddafi saying, "If I fall, Europe will turn black"? I remember I said, "What a weird thing to say!" I didn't realize it, but the Libyan government kept migrants from crossing through Libya.

SONNENBLUME: I had no idea about that either.

KIMBERLEY: And that makes me very sad that these African countries are in such a sad state that people risk being kept as a slave to get out of their homes and get to Europe so they can maybe sweep the streets.

THE MYTH OF "GREATNESS"; NIXON & THE ENVIRONMENT

SONNENBLUME: I feel like we live in this time when everything's getting worse. And I think that when it comes to the climate, well, scientifically it is. But socially, is it worse, or has it always been bad? I mean, from the Native American perspective, it certainly has been bad for 500 years. There was never a time when this country was great, obviously—

KIMBERLEY: 1491 maybe. [*laughs*]

SONNENBLUME: [*laughs*] Right. So it seems like this is a myth that the liberals believe as well: that the United States was a better place once.

KIMBERLEY: Well if they mean there was a time when there was more prosperity for the average worker, yes. Before they deliberately deindustrialized the country and got rid of all the living wage jobs, yes, life was better. And there was an understanding that most people—white people—could have a certain expectation for life. So yes, it was better for them. Ironically, it was better for us... [But] as soon as the liberation movement came, they said, "Oh yeah? You're not going to stay in your place? We can get rid of segregation. We can just put you all in jail." And they've ramped it up at certain points because of the "War on Drugs" or whatever excuse was being used. Black people are worse off. Now we know what a better life means to most white people and what that means to us. We know that's two different things. But more of us were employed. We had better jobs. My family was originally from the Midwest and the generation before me, they did much better. They had good jobs, but now every time I go back to Ohio, it's like, "Well, there was a GM plant but it closed. There was a Frigidaire plant but it closed. There was an A/C Delco plant but that closed." And there's no living wage work. So the other out for black people was always government jobs. And they're getting rid of those too.

So things were better for us because the end of the liberation movement coincided with the neoliberal experiment. So, when people talk about life being better, I think that's what they mean.

SONNENBLUME: They're just talking about that prosperity bubble that happened post-World War II until the end of the 60's, beginning of the 70's.

KIMBERLEY: Right. And it's funny how people describe it. They say, "We had a good job and our wives didn't have to work" and it was very patriarchal. The man brings home the bacon and the woman doesn't have to work. I ask, "What's so terrible about women working?" But anyway, that part of the fantasy I find interesting. So I think that's what people mean about the country being greater, their lives being good. Their individual lot in life was better. Expectations could be higher.

SONNENBLUME: But that same period in time, of course, the 50's and the 60's, the federal government was trying to successfully get rid of a lot of Indian tribes, their legal status as tribes. So a lot of Native Americans lost out at that point or were continuing to lose out at that point. There were controversial policies where they were trying to get people to give up their land. And it was actually Nixon, of all people, who turned that one around. He was a fascinating person.

KIMBERLEY: He was complicated. There were things he proposed, and [Democrat Daniel Patrick] Moynihan worked with him, and they briefly discussed having a—not a minimum wage—

SONNENBLUME: A minimum income.

KIMBERLEY: Yes.

SONNENBLUME: Chomsky has called Nixon the last liberal president. Because he passed the Clean Air Act, the Clean Water Act, the Endangered Species Act. All of these things that are of course now being rolled back. This is one of the things that's really disappointing for me personally. I've spent the last 18 years living on the West Coast and a lot of that time I've been able to go out and spend some time in the forests, and spend some time in the deserts, and this and that. I was involved in the early 2000's with some of the tree-sitting campaigns as an Indymedia activist.

KIMBERLEY: Really?

SONNENBLUME: Yeah, so I got to see that. Being introduced to old growth forests, trees that are hundreds and hundreds of years old and hundreds of feet high, and going up into trees like that. There's nothing like that back here [on the East Coast]. You have to go out there to see it. And it's just amazing. [But] the protections on these places—which have always been tenuous—are being rolled back really fast right now and no one's paying attention. Everyone's just like "Russia! Russia! Russia!" But wait! There's stuff happening to the Endangered Species Act and we're going to have animals and plants go extinct, and that's it for them. That's it.

KIMBERLEY: It's true. What do they say? "There's no Planet B."

SONNENBLUME: Right?

KIMBERLEY: But we forget that all those things [environmental protections like the Endangered Species Act] happened because of activism. I remember the first Earth Day. It was like 1970, I was ten or eleven. A million people protested in Manhattan. They went after what they called the "dirty dozen" members of Congress. That's why Nixon did it. That's why there's an EPA. People made that demand and they [Nixon and his cadre] were afraid for their political survival.

SMART PHONES VS. SOLIDARITY

KIMBERLEY: Now, I feel like several things have happened. People have been numbed out, they've been lied to a lot, so people just don't know. I think technology is to blame, as much as I use it—and sometimes I've said I'm addicted to my cellphone—but having these things can numb you out and distract you at any moment.

Because of technology and the internet you can do things without human help. I was talking to somebody the other day about travel agents. And they said, "I can't remember the last time I went to a travel agent. Maybe in the 90's." Now I can book a trip anywhere on the planet and I don't have to talk to anybody. I can be in my pajamas and slippers with my computer from home. Buy a plane ticket, hotel, etc. Banking: I remember I had a check that I had to deposit and I didn't want to go to the bank. I realized I had an app on my phone and I could deposit the check that way. I've not been to the bank to deposit a check since that day. So I don't need to talk to another human being to do a variety of things. That's another thing that leads us to being cut off from each other.

SONNENBLUME: I think that the smart phone—which was invented eleven years ago now—it seems like there's been this immense cultural shift that's occurred, with no warning. People have just dived into it without any consciousness of it, with very few people paying attention to what it's doing to us, as it's doing it.

KIMBERLEY: Yes. It's meant to be addictive. For example, this morning, I woke up very early and had errands to take care of. But instead I picked up my phone and went on social media. But it's the dopamine in your brain:

click, click, click. And then I had to rush to get here after spending too much time online.

SONNENBLUME: They have psychologists working for them, to make this stuff addictive.

KIMBERLEY: I find that if I really want to not use it, I have to physically put it in another space. I'll be in the living room and I'll say, "Okay, enough with the phone." I have to get up and put it in my bedroom.

SONNENBLUME: But you're thinking about it.

KIMBERLEY: Yes, but if I'm not thinking about it, or if I'm bored—. But if I'm interacting with you, I'm not going to pull my phone out. But I know people who cannot stop even when they're talking to other people.

SONNENBLUME: Yeah, that's strange, isn't it?

KIMBERLEY: Yes, it is. So anyway, we have this country which always, in its history, mitigates against [solidarity]. Although we have had it. We have in the past. We should remember that. People being in solidarity, and acting and working together. But first of all, the movement was crushed. Lest we forget, COINTELPRO crushed the civil rights movement. They murdered the leaders. They put the Panthers in jail. They crushed the Black Panthers.

So we forget that these things were deliberately destroyed and even if you're conscious and want to, you have to be afraid: "What happens to me if I step out? Am I going to prison? Am I unemployable?" So the system works to keep us apart. And so it does take an extra effort to think about how we can work together. And I think with increasing desperation—and the lowering of the standard of living—people [are] less generous of their time with other people. Less likely to think about something else that's not in their immediate sphere.

But it's worldwide, too, and not many people have an international perspective. I mean, look at England. Look at what they're doing to Jeremy Corbyn. He's no Socialist. He's just a not right winger. Tony Blair was just a Bill Clinton of the Labour Party, or maybe an Obama. And he [Corbyn] appears to be serious about truly being a Labour leader and they're destroying him. Everyday there's a different claim that he's an anti-Semite. That seems to be the thing that they've latched onto to demonize him. But that's what's happening all over the world.

THE 2016 ELECTION, THE GREEN PARTY AND MEDIA OWNERSHIP

SONNENBLUME: If you want to know why the Democrats lost in 2016, you could just go and look at what [Greg Palast] uncovered—

KIMBERLEY:—He's been saying it for years—

SONNENBLUME:—Crosscheck and all that. There it is.

KIMBERLEY: Provisional ballots thrown out. And who usually has provisional ballots? Black people. Then they start with the voter suppression and the Crosscheck and all that, but it's not just voter suppression. For example, they say Trump won Michigan by 10,000 votes. There were malfunctioning voting machines in Flint and Detroit—of course that means black people—70,000 votes were not counted. Just weren't counted. So Hillary probably did win. She probably won the electoral vote and the popular vote. Democrats say nothing about it. [Writer] Glen [Ford] calls it a "gentleman's agreement." Because Democrats don't really want to be connected with Black people either. They do need our votes. The way our system is set up, one of the parties is the white people's party, and that's been the Republicans for the last fifty years. So they [the Democrats] know they have us, because we have a well-founded fear of the white people's party being in power. So they don't have to defend us. And the more they are connected with us, the more unpopular they fear they will be with white people. So they don't say anything.

SONNENBLUME: I guess the process of trying to get those 70,000 provisional ballots counted would've been complicated or hard if they wanted to.

KIMBERLEY: And who was it who asked for a count? Jill Stein. And a judge in Michigan said, "You have no standing because you weren't going to win." So if Hillary Clinton would have said something, she could have had standing.

SONNENBLUME: I remember that. I was surprised when Jill asked for the recount.

KIMBERLEY: That was controversial. There were Green Party members who weren't in favor of the recount. They said it wasn't our issue. We're not supposed to care if the Democrats say that we cost them the election. We didn't, by the way, but we're supposed to be building a different party. And I tell people all the time, all this talk of a third party—our system is very hostile to third parties. We need to talk about replacing the Democrats and having a truly progressive party. So I'm not interested in trying to coexist with them. We need to get rid of them. They are the problem.

SONNENBLUME: Right.

KIMBERLEY: They're a bigger problem, in a way, than the Republicans are, because they're phony. They have this veneer of being the progressive party, they're the inclusive party, they're the justice party, or they're the peace party and they're anything but.

SONNENBLUME: And being so smooth-talking about it.

KIMBERLEY: Or just saying, "Look at them! You don't want Trump to be president, do you? Look how racist they are!"

SONNENBLUME: As if toppling Libya wasn't racist.

KIMBERLEY: How about that? And that's why they wanted a black man. That's why they needed Obama. He could get away with it. When I think of the things that he got away with that nobody says anything about, just because Trump is this overt, racist vulgarian, stupid man. But if you know how to use the right knife and fork and talk nice, nobody cares what you do. Or very few people do. They don't say anything.

SONNENBLUME: The same thing happened during the Clinton administration and that's when the prison population exploded and that's when welfare got gutted and that's when the telecommunications act got gutted and led to this crisis in media that we have now where there's—

KIMBERLEY: What is it, six corporations?

SONNENBLUME: About to be five, because Disney's buying Fox. So it's gonna be five [that own over 90% of the media in the USA].

KIMBERLEY: Pretty soon it will be one.

SONNENBLUME: [laughs] I mean, five is one. You know what I mean?

KIMBERLEY: [laughs] It's all the same. It might as well be one.

[Correction: The various news-centered divisions of Fox will be spun off into their own new company and not be owned by Disney. I will quickly add that the "entertainment" divisions of media corporations engage in cultural propaganda that is arguably deeper and ultimately more effective in maintaining the profound sickness of our society. Still, in this conversation we were specifically discussing news media.]

DISSOLUTION AND REVOLUTION

SONNENBLUME: So, does it just go downhill from here?

KIMBERLEY: [pauses] Good question. [pauses again] I think it's hard to predict the future. It's hard to predict what's going to happen. But there's some definite trends we can see. I think the standard of living will continue to diminish for people. Nothing's being done.

Everybody talks about the Paris climate accords that Trump decided he wasn't going to follow but the Paris climate accord didn't do much anyway. So I think our lives will get worse, the state of the planet will get worse, and our lives around the world will get worse.

Unless people rise up. That's always possible. Or, if there's a worldwide war. I mean, I was thinking recently, and I posted on Facebook, would there have been a Russian Revolution without the First World War? Would there have been a Chinese Revolution without World War II?

SONNENBLUME: Interesting.

KIMBERLEY: And I don't know. I know people have pondered these questions. I don't know if you have to have a disaster in order for that sort of mass action to be so successful. And I think maybe you do. I'm not wishing [for] war, because a lot of people are going to get hurt in a lot of ways.

SOCIALISM & VENEZUELA

KIMBERLEY: We are told constantly: socialism doesn't work. So you can keep people from asking about social security, or Medicare systems, or anything that helps people. We're immediately told that it's socialist and that socialism doesn't work. And they use Venezuela. The US destruction of Venezuela: it's all because of sanctions. That's it.

SONNENBLUME: No one wants to say anything good about Venezuela. Everyone wants to be against Venezuela. Everyone.

KIMBERLEY: Because of the lies. People are starving in Venezuela and "it's all Maduro's fault."

SONNENBLUME: Because of our sanctions.

KIMBERLEY: Venezuela has oil it can't sell. I don't know if people know what sanctions mean. So when the United States declares sanctions against a country, that doesn't just mean that the United States isn't going to do business with Venezuela. Anybody who wants to do business with the Unites States, also cannot do business with Venezuela. So they can't renegotiate their debts. You can't send money to Venezuela. If you decided, I'm going to send some money, because I know somebody there and I'm going to help them buy food, you can't.

SONNENBLUME: Wow.

KIMBERLEY: So other countries can't do business with them regardless of how they feel. So of course their economy has cratered. Of course it has. Because it's the US's intent.

SONNENBLUME: It's so sad to see.

KIMBERLEY: It is. But you see the way they talk: "Maduro is Chavez's hand-picked successor." But he was vice president. That's the way our country works. If a president dies, the vice president gets the job.

SONNENBLUME: And he got very high numbers in the election there.

KIMBERLEY: They keep electing him. That's why the US government hates them because people insist on keeping that government. But then they [the US] did something else sneaky: They interfered when Maduro and

56

the opposition were talking to one another. They just want to grind them into the dirt.

US WEALTH

KIMBERLEY: People [in the US] live relatively well because other people are living so badly.

SONNENBLUME: The prosperity that so many people like to glorify comes with a huge price, to people and obviously to the environment... Whenever we pull any of these resources out of the ground, it's a horror show at the site where it's happening. And of course most people never see that kind of thing and they have no idea.

KIMBERLEY: I think there are enough people who are still living somewhat well enough that they are not ready to make change. There's still enough people who are employed. Okay, so you have to get a roommate. Okay, you have to have a job and drive Uber. Both spouses to work. Okay, you have to live with your parents longer. Something. But you still have mobility, you still have an income, there's still enough people who have some kind of health coverage. But when enough people start to lose that, I don't know—it scares me. It just scares me.

SONNENBLUME: Well in 2008, it was bad enough for a lot of people, but they put enough money into the system to keep it from crashing, but they can only do that for so long. There is some point, and I feel like that point could come unpredictably, too.

WAR & IGNORANCE

KIMBERLEY: The only thing the US has is its military power.

SONNENBLUME: Which is still horrifying at this point.

KIMBERLEY: It is.

SONNENBLUME: All those nuclear weapons ... There's so many countries in the US who are like, of course we're not going to mess with the US because they've got the bomb. And we know from history that they're willing to use it. They're completely willing to use it.

KIMBERLEY: And most Americans wouldn't say anything if they did.

SONNENBLUME: No.

KIMBERLEY: It's pretty horrifying.

SONNENBLUME: And again, is that just the national character?

KIMBERLEY: Part of it is. How was the country founded? It was founded on exploitation. It was founded on conquest. During eras of prosperity, segregation was completely legal. Racial segregation was legal. So people got things, the little bit they had, because someone else was being disadvantaged. And when somebody else said, I don't want to be disadvantaged anymore, I mean, somebody else loses something. We have a whole system that has disadvantaged millions of people. Who then want to come here, and we hate them because they want to come here. There wouldn't be any Syrian refugees if the US hadn't decided to destroy Syria. Or Honduras. It's sad.

The thing that scares me the most is ignorance. The lack of knowledge. And people can be ignorant who read the New York Times. Partly they're ignorant because they read [it]. They're ignorant because they watch MSNBC. I think it's funny that people think they're being "active" because they listen to somebody who they agree with. So, Obama's in office so conservatives lose it, so they all watch Fox News. So then Trump wins, and they decide they don't need to watch Fox News anymore, so their ratings drop. And Democrats say, I'm going to watch Rachel Maddow and I'm doing something because I'm watching someone who agrees with me. It's really a substitute for activism, to watch cable news.

CAN THERE BE A HAPPY ENDING?

SONNENBLUME: Considering the fact the US is a settler-colonialist project, is imperialist and white-supremacist, etc., from day one, can there be a happy end to that? Or is it in the DNA of this project that it's just brutal and is going to end brutally?

KIMBERLEY: I'm afraid so. It's not going to end easily or painlessly. It does need to end, though. They do need to get rid of the dominion of the dollar. But that could mean me being unemployed. It means millions of people being unemployed who don't think there's any other way to live, who don't think there's any other system to have, and they'll just turn on each other. [pause] God, I'm sounding so pessimistic. I want to be optimistic about our chances of change.

SONNENBLUME: Well I'm just asking for a realistic appraisal, not for optimism or pessimism.

KIMBERLEY: But optimism can be realistic, too! [laughs] But I do think, to make a long story short: I think so many things are going to get worse. But you've got to be there to tell people the truth.

August 11, 2018

A Century of Theft from Indians by the National Park Service

In early February of 2016, I stopped in the men's room at the Hole-in-the-Wall visitor center in Mojave National Preserve in southern California. Above the urinal was a framed poster about the history of European invasion of the area entitled, "Fort Mojave—*Aha Mocave*—And Then the White Man Came." I was instantly offended by its placement. Who decided that was a respectful way to treat such an important subject?

The poster's voluminous text in three columns, along with eight captioned photographs, presented far more information than could be absorbed in even a long visit to the urinal (not that I'm suggesting it should have been hung in the toilet stall where people spend more time).

The Mojave National Preserve is run by the National Park Service, which in contrast to previous times, has been including more Indian history in its displays and programs, and presumably this oddly-sited poster is part of that effort. Why wasn't it inside the visitor center itself, where one might read it with better attention? And was it also displayed in the women's room? I didn't check. Among the books on sale in the visitor center, one full shelf was dedicated to the topic of Indians, and though the selection was decent, this particular poster—hung on the wall with other ones—would have given the topic visibility to more people.

Getting back into my truck, it occurred to me that someone with a mischievous sense of humor—what some Indians call "coyote spirit"—could have purposely placed the poster above the urinal so that the fact of Indian mistreatment by Europeans would be square in the face of every man who went in there, the majority of whom would be white. (According to research of visitors by the Park Service itself, "ethnic and racial minorities are virtually absent from the major parks in the system," with Blacks and Latinos comprising only 3.5% each, well below their proportion of the general population.[1]) The idea that the poster was there as a prank gave me a wry smile.

Indian history, specifically as it relates to the National Park Service (NPS), had already been on my mind. A few days earlier, I had stopped into a Joshua Tree National Park visitor center to purchase an "America the Beautiful" annual pass, since I planned to travel throughout federal public lands in the West over the next few months and the pass would save money in entrance fees. While there, the ranger pushed into my hand the latest edition of the park's free quarterly newspaper.

The front page story was entitled, "Celebrating the Centennial of the National Park Service." Amidst the expected "rah rah us, we'll be here for another century" promotional banter was a quotation by George Catlin, a famed 19th Century painter of Western landscapes. In an 1833 article in a New York newspaper, Catlin suggested protecting the nation's "pristine beauty and wilderness" by establishing "a nation's park, containing man and beast, in all the wild and freshness of their nature's beauty."

Catlin is credited with coining the term, "National Park" and his vision was of a preserve in which would abound not only plants and animals but also Indians, who were common subjects of his paintings. This view was shared by other European thinkers of the day including John James Audubon, Osborne Russel, and Henry David Thoreau. In 1837, Washington Irving called for "an immense belt of rocky mountains and volcanic plains, several hundred miles in width" that "must ever remain an irreclaimable wilderness, intervening between the abodes of civilization, and affording a last refuge to the Indian."[2] Admiration for the Indians (though often patronizing) was also a theme for many of the Romantics of the 17th, 18th and 19th Centuries. (See also Montaigne's "noble savages.")

These inclusive ideas went out of vogue over the next fifty years. Catlin had made his call before the Gold Rush, the Oregon Trail and the completion of the transcontinental railroad. In a pattern that had already been established in the east, settlers would move onto Indian lands, including areas reserved for the Indians by treaties, and sooner or later bloodshed would begin, more often than not instigated by settlers. The sensationalist media of the time whipped up anti-Indian sentiment. Then the US military would literally "send in the Cavalry" to "protect" the settlers, and kill or round up up all the Indians they could. (These policies have a contemporary analog in the treatment of Palestinians by Israelis, which the US government sponsors. The theater has changed, but not the narrative being played out.)

Meanwhile, in the intellectual circles of the European colonists, new ideas about "Nature" were taking shape. Early conservationists espoused an ideal of "wilderness": land that is "pristine," "untouched" and "unpeopled" that should be protected from industrial activity and human settlement. Considering the rapaciousness of both industry and settlers, they were right in calling for such protection, and we can be grateful today that somebody back then slammed on the brakes, at least in a few places. However, like many intellectuals, they were fundamentally myopic. As Isaac Kantor puts it in "Ethnic Cleansing and America's Creation of National Parks": "The untold story behind our unspoiled views and virgin forests is this: these landscapes were inhabited, their features named, their forests utilized, their plants harvested and animals hunted. Native Americans have a history in our national parks measured in millennia."[3]

These facts were unknown or ignored by most conservationists, who were products of their decidedly Judeo-Christian culture, which gave "dominion" over the entire earth and all its creatures to "Man." Man was apart from nature, not a part of it. "Pure" nature was an Edenic place, free of Man and his inherent sinfulness. Since Man was incapable of having a healthy relationship with nature, the only way to keep nature safe was to keep Man out of it, including Indians. That these ways of relating to nature and to themselves were unhealthy on their own account went unobserved by the conservationists, except among a handful of radicals.

It should be no surprise, then, that when Yellowstone was made the first national park in 1872, the legislation did not set up Catlin's preserve, "containing man and beast." Instead, Congress established, a "public park or pleasure ground of the people," "the people" being European colonial tourists of course. A policy of extirpating Indians from their lands for the sake of preserving "wilderness" commenced.

What follows are examples of the treatment of Indians in the cases of four particular National Parks: Yellowstone, Glacier, Yosemite and Everglades. Afterwards, I will address steps that could be taken to rectify past and present transgressions.

YELLOWSTONE

Upon founding of the park in 1872, the Shoshone who lived in Yellowstone were immediately removed under the terms of an 1868 treaty that the US Congress never ratified. Other tribes, including the Crow, Bannock, Blackfeet, and Nez Perce, who had inhabited Yellowstone seasonally or permanently for generations, saw their hunting rights eroded and then revoked, and finally their access to the park's lands forbidden entirely by 1880, except of course as fee-paying tourists.[4] Park officials falsely claimed the Indian hunting threatened animal populations in the park, though their own surveys showed increasing numbers.[5]

Some Indians still surreptitiously gathered plants and hunted in Yellowstone, taking pains to evade the US Army, which was charged with enforcing the ban. As Kantor tells it:

> *Thus an obsession with halting Indian use of Yellowstone began, despite authorization of much of this use by off-reservation treaty rights. The fact that Native hunters had little to do with the dramatic declines in wildlife populations of the West seems lost on early park management. The incredible irony that early managers were disregarding an integral part of the idealized Eden early white explorers found, its human inhabitants, seems overlooked by people trying to preserve part of what those explorers experienced." He adds, in a note: "Speaking of irony, even as managers struggled to exclude native use, an early park brochure reported that superstitious fear of the park had kept it historically free of the 'red man's yell."[6]*

The Indians' treaty rights—which allowed them to hunt and gather in their historic territories, even when these lay outside their reservations— were nullified through the courts. In July 1895, a Wyoming "lawman" attacked a Bannock camp and stole their property, including some elk meat, and arrested them "for violating Wyoming game laws." A state judge found in favor of the Indians, since their treaty rights preceded the admittance of Wyoming as a state in 1890. However, the Supreme Court overturned the case, ruling that Congress had the right to nullify the previous treaty rights by means of recognizing Wyoming's statehood. In this way, state law was used to deny the Indians their rights on federal land, a true legal perversity.[7]

GLACIER[8]

In 1888, the Blackfeet Reservation in Montana was comprised of its current holdings, plus what would become the eastern half of Glacier Natural Park, a mountainous region called "the Backbone of the World" by the Blackfeet. The area held great spiritual value for them, as well as being hunting and gathering grounds. This reservation had already been greatly reduced in size from its original dimensions, which had taken up two thirds of the state of Montana east of the Continental Divide.

In 1895, suspecting that these mountains contained valuable mineral deposits, the US government sent commissioners to purchase them from the Blackfeet and offered $1 million. The Blackfeet did not want to sell, but finding themselves in dire straits of imminent starvation with winter coming on, they settled for $1.5 million, on the condition that they could continue to hunt, gather and cut timber there. The commissioners agreed to this and the deal was accepted, though with great reluctance by the Blackfeet.

Fifteen years later, the Europeans reneged. In 1910, Glacier National Park was created by Congress, and included the area purchased from the Blackfeet. Since the legislation did not mention the Blackfeet's hunting and gathering rights, legally the original agreement was still in force. But from that time forward, the NPS has acted as if those rights were extinguished, and prevented traditional Blackfeet activity on park land. To this day, the Blackfeet insist that they retain their rights and the situation is unresolved.

Booting out the Blackfeet did not prevent promoters of the national park from using them to attract and entertain tourists, however. Luis Hill, owner of the Great Northern Railroad, employed Blackfeet at the park train depot and in his hotels. The "Glacier Park Indians," as they were called, became well-known as far away as the East Coast. The national park itself used them for promotion, while insisting that they otherwise had no place there. Adding insult to injury, Kantor mentions that Luis Hill "was allowed to commercially fish whitefish, a prize item on menus at his hotels, from park lakes until 1939, when the practice was banned on ecological grounds."

Hill and his Great Northern was not the only railroad involved with the establishment of National Parks:

> Railroad tycoon, Jay Cooke, of North Pacific Railroad, funded expeditions into the lands that would eventually become Yellowstone National Park. On these exploration trips, they brought along people whom they knew would write, paint, or somehow market the beauty of the land, which would therefore increase public and government interest in creating protective laws. North Pacific Railroad saw the creation of a park as an opportunity to profit hugely by extending their lines to the borders of parkland and transporting visitors. The North Pacific Railroad also took the initiative in building nearby hotels for the tourists.... [They] were wildly successful in creating and sustaining a new field of profit and other railroad companies followed suit. The urging of Southern Pacific Railroad helped in the creation of Yosemite National Park.[9]

The railroads had already profited from Indian theft when the US government gifted them with generous rights-of-way from the Missouri to the Pacific. As is well-known, the railroads were instrumental in the hunting to near extinction of the buffalo, an activity meant to deny Great Plains Indians their main source of food and hence starve them off the land. Indian genocide was, then, what would now be called a "public-private partnership."

Yosemite National Park is a special case—the exception that proves the rule—because after the endemic Indian population was driven away, some returned and were allowed to live on park grounds for many years. Originally established as a state reserve by the state of California (after US President Lincoln set it aside into public trust in 1864), Yosemite became a national park in 1890 and was under the control of the Army from 1891 to 1913. During this time, the resident Indians lived in their own village and interacted with tourists, acting as guides and selling them crafts, while trying to hold onto as much of their traditional lifestyle as possible. Other tribes, however, who hunted and gathered in the park but lived elsewhere, were kept out by the Army after 1890, in abrogation of treaty rights.[11]

When the National Park Service was officially promulgated in 1916, Yosemite park staff started up "Indian Field Days," an annual event intended, in their words, to "revive and maintain interest of Indians in their own games and industries, particularly basketry and bead work." The event was intended to draw visitors to Yosemite during the slower summer months, when the park's famous waterfalls had dried up for the season.

At this point in the National Park Service's history, park directors were encouraged to develop their holdings into tourist destinations by providing amenities to attract visitors. This included roads, restrooms,, ranger stations,, campgrounds, lodges, trails, and in the case of Yosemite, a golf course. "Management" of animals and plants to cater to perceived tourist interests was also undertaken, such as killing predator species to encourage game animals like deer, and chopping down trees and vegetation to manufacture scenic vistas. "Indian Field Days" was in the spirit of these development goals and proved a successful draw.

Park officials were unconcerned about accuracy for this event, however. The Indians were instructed to costume themselves as Great Plains Indians and set up tipis, in order to match the popular conceptions of tourists. Their own traditional *umuchas*—conical structures covered with strips of bark—were not considered aesthetic by park officials. They were also paid to participate in parades and to compete in events like rodeos, bareback horse races ("striped as Warriors") and costume contests. In Spence's words, "the Field Days often degenerated into little more than an excuse for tourists and park officials to pose in buckskin and feathered headdress." Indian Field Days continued into the 1920's.

Though the Indians were able to retain some of their traditional lifestyle practices, which included hosting Indians from outside the park for their own events and ceremonies (most prominently the annual "Big

Days"), park officials gradually tightened their control, especially after the final Indian Field Days in 1929. In 1931, the park's superintendent, Charles Thomson, began the process of moving the Indians from their ancestral spot to a new "village" of tiny cabins (only 429 square feet in size, each) for which the residents would have to pay rent. Even with 6-8 people in each cabin, there was insufficient space for everyone who had been living in the park, and Thomson took advantage of this to pick and choose who could move into them. Among those forced to leave were Indians who had generations-long histories on the land. Others were allowed to stay because they were popular with tourists. By 1935, all the Indians who remained in the park had been transplanted into the new cabins.

In the big picture, park officials hoped to rid Yosemite of its resident Indians by attrition, and ultimately they succeeded. Over the next two decades, they raised rents, failed to maintain the infrastructure, and refused to hire first from the Indians, as promised. In 1953, the park instituted the Yosemite Indian Village Housing Policy, which restricted residence to employees and their families. Everyone else had to leave within four weeks. Vacated cabins were destroyed to ensure shrinkage of the community. (This was during the era of the Indian "Termination" laws, which were intended to invalidate Indian sovereignty, break up the reservations, and assimilate Indians into European culture.) In 1969, the remaining Indians were forced into housing for park employees. The following year, the few cabins still standing were destroyed during a firefighting practice exercise. In 1996, the last Indian, Jay Johnson, departed after retiring as a park forester. He had been born in Yosemite.

EVERGLADES[12]

The first agreement between Europeans and Indians in Florida was the Treaty of Walnut Hills in 1793, which recognized the Miccosukee tribe and their rights to territory on the peninsula. Before Florida was even officially ceded to the US, the drive to remove all Indians from the territory commenced, starting with the First Seminole War of 1817-1818, led by General Andrew Jackson. ("Seminole" was a name given to the Miccosukee by Muskogee-speaking Creek Indians who fled south from encroaching Europeans. Their word, "siminoli" means "people of the distant fires" and was borrowed into English to refer to all Indians in Florida.)

In 1830, Andrew Jackson, now president, signed the "Indian Removal Act," which required that all Indians in the southeastern states be sent west, to Arkansas or Oklahoma. Following this, some Indians were "cajoled and threatened" to depart, but others refused. In 1835, the Supreme Court, led by Chief Justice John Marshall, found that, under the terms of the ceding of Florida to the US, the Treaty of Walnut Hills was still valid, giving the Indians every right to stay. This inspired Jackson's famous words, "the Chief Justice has made his law; now let him enforce it" and he continued with his efforts unabated.

In 1842, Colonel William Worth declared "peace" in the Seminole Wars, and under the terms of the Macomb-Worth Agreement, granted the Indians the southwestern corner of Florida, including what would eventually become Everglades National Park. In 1845, President Polk amended the agreement with an executive order reserving a 20-mile wide strip around the entire area that was to "reserved from survey and sale."

Nevertheless, the Seminole Wars continued until 1855 and the Macomb-Worth Agreement was largely ignored by settlers and the military. However, the US never fully accomplished its goal of driving out all the Indians, despite great monetary expense and high casualties. Though the majority of the Indians were sent west, a few hundred tenacious individuals—mostly Miccosukee but also some Muskogee—managed to hold out, hiding in the wetlands of the Everglades, which they call, "Pa-hay-okee." There, they refused offers of money for land and in their isolation succeeded in avoiding the poverty and cultural dissolution that had been suffered by Indians elsewhere who were shunted onto reservations far from their homelands. They made no agreements with either the federal or state government.

In 1935, Secretary of the Interior Harold Ickes and Indian Commissioner John Collier met with some of the Indians, mostly Creeks, to try to convince them to become an "official" tribe and seek a legal reservation. Ickes also informed them about the proposed Everglades National Park, which would take a significant piece of their traditional (and treaty-guaranteed) territory, though Ickes reassured them that he believed they should retain hunting and gathering rights there, and have first shot at park jobs.

The majority of the Miccosukee had no interest in giving up any land or in negotiating with the US, but a group of Muskogee felt differently, and went to the table with the Bureau of Indian Affairs and an above-board reservation was created for them which was declared the official "Seminole" reservation. Starting in 1937, the government began forcing Indians out of the Everglades, and in 1947, Everglades National Park was opened. This outcome, and the method of arriving at it, was typical: splits within Indian populations were exploited by the US, which would ignore traditional factions in favor of "progressive" ones. Traditionals, standing by their ancient principles, would be left out in the cold, knowingly. In this case, the "progressives" were dominated by a non-endemic tribe, the Muskogee

Creek, and the Miccosukee traditionals were not considered to be a "real" tribe by the US.

In 1946, the Indian Claims Commission (ICC) was formed, with the ostensible purpose of settling any extant claims with cash payouts. The real purpose was not justice, but to open Indian lands to resource exploitation. Any tribe that accepted a settlement would give up previously held treaty rights. In 1950, the Muskogee progressives filed suit to receive payment but the Miccosukee traditionals loudly distanced themselves from the effort and in 1951 presented President Eisenhower with "the Buckskin Declaration," which demanded that the US cease conflating them with the Muskogee, respect their long-standing independence, rights and lifestyle, and send a representative so that these issues could be cleared up.

Eisenhower sent the US Indian Commissioner, who concluded that the Miccosukee had the right to live and hunt in their traditional ways in the National Park. In 1954, in a report following a meeting with the Everglades Miccosukee General Council, the US finally recognized the Miccosukee as a separate tribe from the Muskogee and reiterated that their rights of occupancy and use still existed, regardless of claims to the contrary. However, the ICC ignored this report, and the NPS refused to allow Indians to hunt or live on park lands. The theft was admitted but then quickly ignored.

Rifts formed among the non-acknowledged Miccosukee during the 1950's. In 1960, the General Council—without the support or approval of a group of traditionals—successfully forced the state of Florida to recognize them as a tribe and extend them a lease on 143,500 acres of land. The Bureau of Indian Affairs soon granted them a lease from the NPS for a "reservation" five miles long and 500 feet wide along the Tamiami Trail, an east-west highway north of the National Park. The opposing traditionals refused to cooperate in this process and did not join the "Miccosukee Tribe of Indians of Florida" when it was incorporated in 1962. In 1976, the incorporated entity accepted a cash settlement from the ICC.

In 1974, the NPS swallowed up another chunk of ancestral Indian territory with the creation of the Big Cypress National Preserve. Though the terms of the theft allowed the Indians to retain "their usual and customary use and occupancy" there, including hunting, gathering and conducting ceremonies, the NPS proved reluctant to honor the agreement, especially "occupancy"—which is to say, tending their gardens and actually living there.

In 1983, President Reagan signed into law an agreement between the incorporated Miccosukee tribe and the state of Florida granting the tribe 189,000 acres and $975,000. Shortly thereafter, he received a letter from some of the traditionals—the last hold-outs of the last unconquered Indian tribe in the US—protesting the agreement as illegal.

2015: The NPS Tries to Loosen Up (a little)

In 2015, the NPS proposed a new rule that would allow traditional plant gathering on park lands by Indians, with restrictions. In its text, the NPS takes a step away from the hallowed concept of "unpeopled" wilderness:

"Research has shown that traditional gathering, when done with traditional methods and in traditionally established quantities, does not impair the ability to conserve plant communities and can help to conserve them, thus supporting the NPS conclusion that cooperation with Indian tribes in the management of plant resources is consistent with the preservation of national park lands for all American people."[13]

In Yosemite, after the resident Indians were no longer allowed to follow their traditional practices, which included controlled burns as an important activity, their gardens became overtaken first by shrubbery and then trees, and were no longer productive for food or medicine. The proposed rule from the NPS does not go so far as to allow burning as a wild-tending technique and is silent on the topic of seeding or other forms of propagation. So either the full picture is not clear to them yet, or they are only interested in taking this one small step.

The proposed rule comes with a number of restrictions:

- Harvesting can only be done by members of federally-recognized tribes with historical ties to the particular park land;
- Names of harvesters must be submitted along with an explanation of how they were chosen;
- A list of what will be harvested, where and how much, must be included;
- The park superintendent or regional manager may deny the activity based on environmental grounds.

To bureaucrats these restrictions make sense, as they probably do to many people who are accustomed to accrediting themselves with bureaucracies and for whom such obedient acquiescence (if accompanied by defeated grumbling) is a mark of responsible adulthood. But traditional plant gathering has never been subject to such processes. If anything, the ethos that guided traditional activity was far stricter, though in a way that is virtually incomprehensible to the colonial mindset. It's not for nothing that Indian gardens were productive for many centuries, in sharp contrast to the farming methods of the European settlers, who moved on after they had worn out the soil, often within one human generation or less.

As per their process, the NPS invited comment to the proposed rule, and received 69 responses from Indian tribes and individuals. As reported by the Indian Country Today Media Network (ICTMN), the rule's specifics did not meet with a chorus of approval:

> *Most who submitted comments had something to say about the authority vested in the park superintendents, the designated gatherers and limiting them to tribal members, and the disclosure of traditional information with no assurance that this information won't be made public (sharing these details with anyone outside of the tribe is not customary—period).*[14]

As seen in the section above on Everglades National Park, not all Indians choose to be "federally recognized" but still claim their legal rights to hunt and gather in their traditional territories, rights that still technically exist according to treaties, whether or not the NPS recognizes them. As a case in point, the ICTMN article quotes Bobby Cypress Billie, of the the Original Miccosukee Simanolee Nation, who said, "Even if they come out with it, we are still going to maintain our way of life and our land." Other non-recognized Indians might be more easily cowed, of course, due to their own particular circumstances or history.

In some cases, tribes and individuals are not "federally recognized" even though they would like to be. The BIA is, and always has been, a politicized organization, acting not only from its own prejudices but also at the behest of outsiders with power who have their own agendas. There are still tribes, especially in California, who were "terminated" in the 1950's and 1960's, and are still striving to be re-recognized. For the time being, these Indians would not be eligible to participate in the new NPS process.

Just how much distance lies between the worldviews of the NPS and traditionally-connected Indians was made clear by the words of the Indigenous Elders and Medicine People Council, who were quoted in the ICTMN article: "Our sacred relationship with these plants, and the ceremonies and customs connected to them, is an essential part of our spiritual way of life.... To place the Superintendent and/or Regional Director and/or designated person between the Creator and the Indigenous Peoples is a direct violation of our rights."

Again, for those people who are deeply ensconced in the colonial culture of the US, with its endless number of "designated" persons whose permission or grace is required for any number of legal or moral choices, the statement of the Indigenous Elders is ludicrous, even offensive. As a commenter to the ICTMN article put it: "The rights of wild plants and animals MUST take priority over the alleged 'rights' of humans—ANY humans. It's the only way that irreplaceable genetic diversity can be preserved! Wildlife can't protect themselves from us, and extinction is forever!" These words are cast from the same mold as those of the 19th Century intellectuals with their "unpeopled" wilderness ideal, and though they are embellished with current scientific terminology and a contemporary catchphrase, at their base is the same unproductive misanthropy.

At the heart of these misunderstandings is the refusal of European colonial culture to recognize that its view of the world—as "dominion" to be exploited or "protected"—is not shared by traditional Indian culture. No real conversation can happen while the this mindset itself is non-negotiable.

Cutting to the chase, Sherri Mitchell, a Penobscot Indian, attorney and director of the Land Peace Foundation, is quoted in the ICTMN article as saying, "The only appropriate step here would be to remove the prohibitions placed on indigenous people completely, not create new rules that allow restricted access that actually defeats traditional purposes."

Responding to an email I sent in late February, Jay Calhoun, a NPS Regulations Program Specialist, said that the NPS "is considering the public comments that it received" and "hope(s) to publish a final rule this summer."

THE BIG PICTURE

Of course, this proposed rule is small potatoes (or yampah roots) in the larger issues around NPS lands and Indians. For a wider perspective, I share here the words of Peter Matthiessen, from his book, "Indian Country":

> How just, how right it would be... to restore to the Miccosukee of all factions their traditional use of Pa-hay-okee. Excepting the main access route far to the south, the Everglades National Park is almost roadless; in the region of the Long River where the Miccosukee lived before, there are no roads at all. Plenty of territory exists for the few families who might wish to return to traditional life: the tourists would never see them, never hear them. For the rest, the knowledge that there was a homeland for their children to return to would suffice... Since no white people use this north half of the park, and since it is legally Indian land, why not restore it to the Indians? This is their country, and the Indians would certainly observe the restrictions necessary to protect rare plants and animals; they have done so forever, needing no laws, out of reverence for their native earth and sacred homeland.[15]

A still bigger idea is to simply return most if not all public lands to the Indians, along with unconditional monetary recompense if they want it, and full apologies to make it official. This concept is only entertained at the radical fringes of these discussions, but it should be our starting point. After all, it was only through intimidation, violence and theft that Indians don't have access to their land in the first place. To use the language of the religion of the colonizers: if we are looking for redemption, shouldn't we start with the original sin?

When people with typical colonial mindsets discuss Indian rights to hunt, fish, gather and wildtend on public lands, they typically jump right to "what could go wrong." Thus, the restrictive approach of the NPS in its proposed rule and the shrill insistence by the ICTMN commenter that the worst case scenario is the only possible result. But what if we turn the question around and ask: "What could go right?"

The NPS itself broaches this topic in the text of its proposed rule, stating that it "would provide opportunities for tribal youth, the National Park Service, and the public to understand tribal traditions."[16] Understanding of tribal traditions is certainly in short supply these days, not only among Europeans but also many (if not most) Indians. Though recent books and scholarship—generally by non-Indian researchers—have begun to examine how Indians were essentially the keystone species in many ecosystems in the US prior to the European Invasion, it is indeed true that only through actual practice can Indian traditions be kept alive and their vital lessons spread and passed on.

Left to their own devices, unencumbered, is the only way that Indians will be able to keep their traditions alive and amend them as needed for the changed circumstances to the land in their already too-long absence from it. In the past, their traditions would not have been fixed; like everything else in nature, their actions would have been fluid, and it is important that we see what novel ideas and practices would emerge now. European domination of the Americas has resulted in widespread environmental destruction and if there is any chance at all for the survival of all humans and the rapidly disappearing flora and fauna of the planet, we must learn and apply all that we can of "the old ways" Why? Because the old ways worked.

Once Indians had the space to be themselves again, would they be willing to share with us non-Indians? I'm sure that some would and some would not. The topic is definitely controversial in Indian circles. For example, Matthiessen quotes Buffalo Tiger, of the Miccosukee Tribe of Indians of Florida, as saying: "We survive, because we go with nature, we can bend, we are still attached to the earth. Now your way of life is no longer working, and so you are interested in our way. But if we tell you our way, then it will be polluted, we will have no medicine, and we will be destroyed as well as you."[17] Conversely, other Indians have the opposite opinion, and have shared their practices and ceremonies with non-Indians because they were to told to do so, either by their elders or by their own personal conscience. We would just have to see who was willing to share what.

We European colonizers on this continent are like spoiled-rotten children and the work we must do to attain actual adult responsibility is not limited to what Indians might have to teach us. Regardless, it's well past time that we got out of their way.

March 7, 2016

Addendum

"Gathering of Certain Plants or Plant Parts by Federally Recognized Indian Tribes for Traditional Purposes" took effect as a new National Park Service rule on August 11, 2016.[18]

NOTES:

[1] Ruppert, David. "Rethinking Ethnography in the National Park Service" in The George Wright Forum (Vol. 26, No. 3, 2009), p. 53.

[2] Kantor, Isaac. "Ethnic Cleansing and America's Creation of National Parks" in Public Land & Resources Law Review (Vol. 28, 2007), p. 45.

[3] Kantor, p. 42.

[4] Vernizzi, Emily A. "The Establishment of the United States National Parks and the Eviction of Indigenous People" (Senior Project, California Polytechnic University, 2011), pp. 29-30, 33.

[5] "American Indians in National Parks" at Inforefuge, p. 3.

[6] Kantor, p. 50.

[7] Kantor, pp. 50-51.

[8] Information on Glacier, except where otherwise cited, from Kantor, pp. 51-52.

[9] Vernizzi, pp. 22-23.

[10] Information on Yosemite, except where otherwise cited, from: Spence, Mark."Dispossessing the Wilderness: Indian Removal and the Making of the National Parks" (Oxford: Oxford University Press, 1999), Chapter 8: "Yosemite Indians and the National Park Ideal: 1916-1969," pp. 115-132

[11] Vernizzi, pp. 20-21, 27-28.

[12] Information on Everglades mostly from: Matthiessen, Peter. "Indian Country" (New York: Viking Press, 1984), Chapter 2: "The Long River," pp. 15-63; with details from: LaDuke, Winona. "All Our Relations: Native Struggles for Land and Life" (Cambridge, MA: South End Press, 1999), Chapter 2: "Seminoles: At the Heart of the Everglades," pp. 24-45.

[13] National Park Service. "Gathering of Certain Plants or Plant Parts by Feder-ally Recognized Indian Tribes for Traditional Purposes" (Federal Register, 4/20/15), p.6.

[14] Tirado, Michelle. "National Park Service Does Face-Plant With New Rule on Gathering Plants" on Indian Country Today Media Network, 8/20/15.

[15] Matthiessen, p. 61.

[16] National Park Service, p. 2.

[17] Matthiessen, p. 36.

[18] Federal Register. "Gathering of Certain Plants or Plant Parts by Federally Recognized Indian Tribes for Traditional Purposes." https://www.federalreg-ister.gov/documents/2016/07/12/2016-16434/gathering-of-certain-plants-or-plant-parts-by-federally-recognized-indian-tribes-for-traditional

Biomass Is Not So "Green"— Getting the Facts from Josh Schlossberg

Josh Schlossberg is an investigative journalist, horror author and former environmental organizer who lives in Denver, Colorado. He is also the editor-in-chief of the Biomass Monitor, a subscription-supported publication that bills itself as "the nation's leading publication investigating the whole story on bioenergy, biomass, and biofuels." In early September 2018, I was visiting Colorado and met up with Josh. We talked biomass, "renewable" energy, wildfires, politics and activism. What follows is a transcript, edited for clarity. This is an expanded from the version that was originally posted online, which ended at "I don't know what else to do."

KOLLIBRI: This term, "biomass"—I'd just like to start there. I think a lot of people hear that word and they're like, "Oh it's got 'bio' in it, this is something good that's going on," and they move on. But it's a more complicated issue. So what's included under this term, "biomass?"

JOSH: A broader term would be "bioenergy." It all depends on how you want to define things, but "bioenergy" itself would include liquid fuels made of plants or various materials like that. Most biomass is trees for electricity or industrial heating. That's typically what people mean when they're referring to that. Trash incineration sort of falls under the category of biomass. The burning of black liquor which is a byproduct of making paper. Even wood heating, in a sense, is biomass. Where people are most concerned is around forest biomass for electricity and/or industrial scale heating. Because that ties into specific ecosystems stuff.

I don't have the figures off the top of my head right now, but basically half of, quote, renewable energy is a form of bioenergy. I believe about 10% is for electricity. So it's not even the lion's share of what is bioenergy, but it's most of what people are paying attention to. If you're doing biomass, presumably for climate issues, and then you have forests that are doing a lot of work for climate on their own, that's kind of where a lot of the controversy comes into play. And it goes further than that. There's also health concerns.

People are like, "let's get off of coal" because of climate stuff and people are concerned about air pollution. But biomass is also burning stuff. If you are concerned about the emissions from a coal plant, you should be concerned about biomass. Obviously there's lots of filtration. They filter out the great majority of pollutants, but not all of the pollutants.

It actually takes more biomass to generate the same amount of electricity. So you have a chunk of dense coal—and this is certainly not an endorsement of coal, it's just scientific facts—and it provides more energy because it's dense. I don't have the numbers, but you need way more wood because it's less dense.

KOLLIBRI: Volumetrically a larger amount?

JOSH: Yes, you get it. So that burning of more material is going to create more pollutants and particulates for that same amount of energy. So it gets, in some ways, complicated really fast. At the same time, there's some pretty elementary science. If you let a tree grow it continues to absorb and store carbon. If you cut it down and burn it, it releases that pulse immediately.

Industry's argument is that eventually that tree is going to burn up in a wildfire or it's just gonna die in 100 years and release the carbon, so why not cut it down and make use of it? Of course, the issue is about time frame, right? It's about that one pulse that will go up in the atmosphere versus gradually being released over a period of decades because the tree doesn't just evaporate.

Even when it burns in a [wildfire], most of the material is still there and it's stored in the trunk itself and in the roots and in the soil. Then when it falls, it has all these other purposes. A dead tree, even if it did burn in a wildfire, has all these purposes. Then you're getting out of just the Climate Change argument and into more of an ecosystem argument. So its gets, like I said, fairly complex in that regard.

But again, the basic science is they're storing carbon dioxide, so wouldn't we want more forests? Even the industry acknowledges that there are limits and that there are dangers in burning up our carbon storage.

KOLLIBRI: So in the United States right now, where are the hotspots for this?

JOSH: That would be California, the Northeast—primarily Maine—and there's a fair amount going on in the Southeast as well, around Georgia, Florida, Texas. Then they're scattered around the country. There's a little bit in Arizona, one here in Colorado that I've investigated a fair amount and even gone out to the forest there. I'm the only person I know who's taken pictures of the logging operations for that. Frankly, at almost any biomass facility, most people haven't done that. I've tried to document—not even having an opinion on it—here's what logging for biomass looks like.

But the areas in which they're looking to expand are more also in the Northwest and the Southeast. They're the areas that have the most likely potential to expand. It's expensive and so it's really based a lot on subsidies. So [there's] been all this to-do up in New Hampshire and Maine recently with subsides and with facilities closing down.

That's the other thing. Lots of facilities have closed down over the last five years. There was a boom around 2010, partially because of the stimulus package, so a lot of money went toward biomass facilities and tax subsidies and stuff like that. Then natural gas got cheap—temporarily because it's a bubble like anything else—so a lot of these facilities shut down, completely closed their doors, or they sat idle for a while. Some of them transitioned over to natural gas. So the industry was not doing particularly well recently.

However, the EPA had been discussing whether or not they should account for carbon dioxide, were there to be some sort of carbon tax or whatever. They've been discussing this for eight years, I believe. Finally, there was an announcement made saying that they would not account for the carbon dioxide. And it is not a rule as of yet—it would have to get through a rule-making process—but it's a policy directive, and it's basically going to guide their future policy.

So basically they're saying, if you cut down a tree and you burn it up from a "managed" forest, those trees will count as if it's *not* carbon dioxide. As if it's a windmill turning around. And some people think that's legit. I think it's worthwhile for people to genuinely try to understand how industry is looking at the science.

There are hundreds of foresters with academic backgrounds that believe in that concept of biogenic carbon. So the idea that this tree carbon is going up in the atmosphere anyway—it's only temporarily here—all this is just in this ongoing pool, so we can access it instead of coal. [Coal] is in the ground, we take it out, and it's a pulse that would never have gotten out there. So they distinguish between those two.

The thing is—and a lot of scientists say this—the atmosphere doesn't care where your carbon comes from.

KOLLIBRI: Right.

JOSH: It doesn't distinguish that. So if we're at a point where we're at carbon saturation or whatever you want to call it, more of a pulse is not necessarily a good thing. And let's say we had a hundred years. Because that's what they're saying. We'll look at it over a hundred year time frame. More trees would be absorbing that carbon over time, so it's all "carbon neutral," as they say.

But that's not necessarily true. Sometimes forests don't grow back at the same rate, sometimes they don't grow back at all, sometimes they're cut down and developed. There's soil fertility issues. Climate Change is changing the zones. All that can happen. But let's just say all the forest will come back and reabsorb that carbon over a hundred year time frame. Do we have a hundred years to reduce the emissions? So that's really encompassing the argument as best as I can understand it.

KOLLIBRI: When you mention these different places where it's going to be active in the United States, or is active, that's mostly with trees?

JOSH: Yeah, the vast majority of it for the biomass power facilities and the heating facilities is with trees. liquid fuels are primarily from corn and canola and they're working on some other stuff. They do want to do liquid fuels from other things and from trees as well. There's a facility that just opened in Oregon called Red Rock Biofuel and they want to make liquefied fuel out of trees, but the process hasn't really been perfected so I don't know exactly what they're doing.

KOLLIBRI: Not all of the trees being cut down are for US use. Some of it is for European use.

JOSH: That's true in the Southeast. They're cutting private forest land and industrial forest land and they're shipping that overseas. It's a smaller percentage, though, than what's going on in this country. It's interesting, there's more attention being given to the exporting of those trees than to domestic production of it. Of course it's because, "Why do they get our stuff?" and that's a valid concern but there's still plenty going on in the US.

On the West Coast they're actually starting to ramp up for more exports to Asia. So Asia is looking at more biomass. Japan—because of their experiences with Fukushima—they're looking at doing other things. So there's going to be more and more exports. So even it's not as financially viable here, they may export overseas.

It's not like, here's a forest, they clear-cut everything and use everything from the trunk to the leaves and that goes to biomass. It's only sometimes that easy. That's what makes this issue more complicated. I actually documented this with the Eagle Valley Clean Energy facility in Gypsum. That's about two hours west of here on I-70. You can see if from the road. They've been harvesting beetle-killed forest up in the Rockies and clearcutting hundreds of acres and using every bit of the tree for that biomass facility.

That's rare. Most of what happens is that there's a logging operation that's typically already planned and they're using the merchantable wood —typically the lower part, most of the trunk—for dimensional lumber, things like that. Typically the top parts that are more tapering and whatnot are often what go to biomass, as well as any tree that's crooked, knotty, partially rotten, has imperfections. So they will cut those down and use the entirety of those. So it's called "whole-tree" logging but it's a been a bit misrepresented, even by some activist groups who are saying, "Oh, whole tree logging means they're cutting down the whole tree down and using it for biomass." Not always the case. It means they're taking the whole tree out of the forest, which is more ecologically detrimental than leaving the tops and branches, which have the higher nutrient content. So in typical logging just for lumber, they would leave that.

KOLLIBRI: The "slash"?

JOSH: Yeah, the slash. So you know about that. But sometimes they're taking all of it out and a lot of times, biomass piggybacks on these existing logging operations. But in some cases it's entirely just for biomass. And in other cases, it makes the logging economically viable to begin with. So even if all of it is not for biomass, the fact that they have that extra chunk of change for biomass might make it so more of these logging operations happen.

Then a lot of what's happening now is wildfire "fuel reduction" projects out in the forest, throughout the West in the National Forests. That's kind of the big thing that the Forest Service is going for and various places, particularly California and the Northwest, a few of the areas in the Southwest, are going for that. That's a big to-do as well.

KOLLIBRI: And that's highly controversial.

JOSH: It is.

KOLLIBRI: Because there are definitely scientists who don't think that really works for reducing wildfires.

JOSH: Yeah, so there's a lot of science on that. I've focused a lot on that element. And that's quite complicated in many ways as well. You know, the conventional wisdom, is, "Well, there's too many trees 'cause they grew up and we suppressed wildfire, so we need to log them to put things in order."

Now, it's a little more complicated than that, of course. We did suppress wildfires over the last 75 years but there's a lot of scientist suggesting that fuel levels alone are only a small component of what creates a large wildfire. Mostly it's climate. So when there's drought, heat and wind. Anytime there's a big fire, pretty much—I don't want to say 100% of the time, but almost 100% of the time—when you see a big wildfire happening, it'll be hot as hell, super dry for a while, and there will be heavy winds. And when it stops, it's because those things stopped.

KOLLIBRI: Firefighters will tell you that, too.

JOSH: Yeah, they know.

KOLLIBRI: The fire stops when the rains start. They're out there to triage, to mitigate in particular places—keep this house from burning down or whatever.

JOSH: To keep it from going too far. Now, in the backcountry, there's never been anything they can do. And then there's a lot of science showing that it stops some of the smaller fires which are not necessarily what people are concerned with for burning their homes down.

So what most people are saying who really look at the science is: protect homes, the areas around the homes. So it could be that 60 feet around a home is all you need to maintain, to keep the home from burning.

KOLLIBRI: They say 100 feet in northern California.

JOSH: Yeah. Newer studies are saying it's not even necessary to have to go that far but the idea is you do work around the home.

With the beetle-killed forests, it doesn't look pretty. But basically forests have been full of bugs and fires forever. It's not only okay, it's essential. The real issue is when it comes to communities. That's for real.

Of course, people keep building in the forest, so there are issues with zoning. But that's almost more controversial than biomass. The minute you start saying where people can and cannot build their home, people don't want to hear that. But that's the reality: you build your place out there, it's more likely to burn, so there should be more attention paid to that defensible space. That fire-wise stuff. I'm a big advocate for that. Even people who think biomass is great and we should cut down all the back country, still agree we can do something around the home.

But there isn't a lot of science that clear-cutting or even thinning black country native forest will protect anybody. And in fact, it can make things worse by opening things up to more sunlight, drying it out. There are some studies that have shown that in some cases, it has slowed down some of the smaller fires, but not the bigger fires. So is it even worth spending money on it, is the question.

KOLLIBRI: I started studying the forest issues in the early 2000's, when I first moved to Portland, because I was hanging out with tree-sitters out there. I got a lot of education about this and about fire ecology and it is simply a fact that the profit motive leads to all sorts of assertions. Know what I mean? Corporate media, who are at least in the same class with—if not in interlocking boards of directors with—these kinds of companies, are of course going to deliver particular kinds of messages. So the reality around the science of forests and specifically of fire ecology just doesn't actually get out there.

JOSH: It's been a little more. There's some folks who've have been getting it out there. Media has occasionally been paying attention to Chad Hanson.

KOLLIBRI: Yeah, he's really good. At the John Muir Project.

JOSH: Yeah, and folks in Oregon like Dominick DellaSala. And there's lots of folks in the grass roots who have been saying this stuff forever and occasionally people pay attention to that.

But yeah, with biomass, with the wedding of biomass of wildfires, industry can say, "We have to treat wildfire" and "Look, we get our byproduct of biomass so it's a win-win and we're protecting the climate—

KOLLIBRI:—and it's "renewable," la la—

JOSH:—so it's a good story and there's some kernels of truth to it but I think when you look closer at the science, you see that it's opening a Pandora's box, depending on our forest for that amount of energy. It's only a small percent anyway. I mean, right now, I think it's like 1% of our energy so let's say we double or triple. Ooh, 2% or 3%! It's almost an insignificant amount of energy.

So at best it's a distraction. Let's just say an angel gets its wings every time you cut down a tree and burn it. Still it's a tiny percentage of our electricity so we need to be figuring out other stuff. I think the really deep change—and that's the big danger of my advocacy for renewable energy— is you're not looking at our energy consumption. There's energy efficiency, there's conservation, like turn off the lights. That's great. But living more simply and just really reassessing our economic system, but no one wants to talk about that. But that's where I try to take the conversation. Alright, biomass, blah blah blah, but what about how we are living? What about how our cities are set up? All those sorts of things are the ultimate conversation that needs to be had.

And I think even a lot of environmentalists are like, you put in renewable and you get on with your day and everything is good. Democrats have actually been the biggest advocates for biomass. The biggest advocate in Congress is Senator Wyden.

KOLLIBRI: Oh really? Wyden? I didn't know that.

JOSH: He's huge. He's been the biggest advocate.

KOLLIBRI: Because it's "jobs?"

JOSH: It's jobs and it's "renewable" energy. That's where that whole left-right polarity becomes a problem. And also because it's a kind of surface-level environmentalism. Oh, it's "renewable!" Well, we're just talking about climate stuff and you don't like coal, but this stuff actually generates more —depends on how you looks at biogenic carbon—but if you look at it the way more scientists do, it generates more CO_2 than coal.

So it's weird to have folks who are with some of the big climate movements who are saying "no" to coal but then "yes" to biomass energy, which actually produces a lot of CO_2. So it's just inconsistent.

KOLLIBRI: It's wishful thinking, too.

JOSH: So that's what's so complex about it. The environmental community is a bit conflicted on it and more recently they've been at least coming out against industrial-scale biomass electricity. But there are actually a lot of those folks who are even quoted as opposing that [but who are] some of the bigger advocates for other forms of bioenergy. And I'm not putting moral judgments on any of that. I'm just staying that there's a lot of inconsistency and it's very confusing and there's not a clear message being sent.

In the *Biomass Monitor*, it started as an advocacy publication, but I've changed it to getting all sides of the story there. And you know, I have my little editorial, and I will talk about things this way, basing it off the science. But industry has some points to make and I think they deserve to be in that conversation. So what I want to do is see that conversation happening in the same place. So with my articles I will talk to everyone and I will do these point/counterpoint segments. I've had CEOs of bioenergy industries and I'll get someone to write on the other side of it.

Media—mainstream media—is supposed to do that, but they suck at at it. They don't care or know enough or have the time or the resources to delve in to that. Media is not doing that so I try to do that in my tiny-ass publication.

KOLLIBRI: Media talks about not having resources but their resources are enormous. They could cover any of it if they wanted to. But you have 90% of them owned by 5 or 6 corporations so there's lots of things they're not going to cover. NBC is owned by GE right? So did they ever talk about how the Fukushima reactor is a General Electric design? That sort of thing is never going to happen. Or how many of that same design there are in the United States right now. [Note: NBC News is no longer owned by General Electric, but was in 2011 at the time of the Fukushima disaster.]

JOSH: Some of the journalists have like a three day deadline and they don't have an in-depth understanding of the science so they're just going to like, "What does the NRDC say?" and "What does that person say?" "I guess we'll never know." Well, we could look at the science!

But I still think the discussion is important and sometimes some conclusions can be made around that. And sometimes the activists are not making honest statements either and I don't think that really helps the cause.

KOLLIBRI: Well, they get involved in the politics of it. And the level of politics is not about facts, it's about emotion. So, what are the words that are going to trigger certain things? That are going to get people on my side versus their side? Etc., etc. This is what activists do. I've been in that position of being tempted to communicate that way. But activists can never win that game against these corporations that have such larger budgets.

JOSH: No, you can't beat them at that. You can only beat them at the integrity game. That's what is really important for activists. Maybe you trick people in the short term, but then you lose your credibility and that's all that activism has: the credibility of the moral high ground. Once you lose that, you are not going to win. You can't out-propaganda them. They're going to win.

KOLLIBRI: How did biomass fit into Obama's "all of the above" energy policy?

JOSH: He advocated for it a lot. I wrote an article when [he ran against Romney], and there were all these opposing things, but their biomass position was identical.

Biomass was a big thing for the Democrats because it was low hanging fruit for renewable energy. You need a shitload of windmills to produce a lot of electricity, and solar panels. But biomass is an easy way to get in there. That's why they advocated for a lot of that.

KOLLIBRI: There's a mostly untold story about what happened in the Obama administration with energy policy and how the fossil fuel production in the US grew to a level it hasn't been at since the early 70's. When he said "all of the above" he really meant all of the above. So fossil fuel production went through the roof. Along with that the, quote, renewable part was subsidies or support for the industrial-sized plants like the industrial-sized solar and wind out in the desert—

JOSH:—which have ecological impacts—

KOLLIBRI:—which have a tremendous ecological impact. You've provably heard of Basin and Range Watch?

JOSH: Yeah.

KOLLIBRI: I feel like they're the folks to go to first on this one. I got to interview them a couple years ago and learned about that. [See "Taking on the Sacred Cow of Big 'Green' Energy," earlier in this volume] The deserts of southern California and Nevada there, because nothing else was there, they are some of the closest to pristine or untouched areas left in the US. Nothing to chop down, nothing to dig up—

JOSH:—no one wants to live there—

KOLLIBRI: Right? And now they're out there bulldozing all of it to build these massive operations. When actually what should be happening is this should be happening at site. Put solar panels on the roof of every building in Las Vegas rather than—

JOSH: Yeah, that's industrialized rather than more of a localized, smaller scale. We dealt with some of that in Vermont as well because there were ridgeline wind towers they wanted to put up and there were folks advocating against that because forests were being cut. Which was ironic because the people advocating against cutting down the trees on the ridge were actually advocating for biomass power.

It's a whole big confusing mess. The environmental movement has never been on the same page and I think that's part of it's failure. I like to point that stuff out and I'm certainly pro-environment but I want to see the sides do their jobs. Industry's job is to make money and to advocate for everything to do to make money. You can't expect them to do anything else.

So the countervailing force is the activist movement. But there is no real cohesive push-back so industry gets away with stuff. And industry is almost like, "We do what we get away with and you folks don't actually have any solidarity against us so what are we supposed to do?" It's like an alcoholic. They're addicted and they're not the ones who are going to stop themselves. So if the environmental movement can't even get its shit together to have a coherent platform then it's like it's equally their fault.

I say that as a former activist. I was an activist for twelve years and I don't consider myself an activist anymore. I just write about it and I have my own feelings about it. I find my own feelings are not that interesting. My own opinions are not that interesting. I think I can do more of a service [as a journalist] because I can understand the different points of view.

So what does it look like to be a logger in Maine, and this is all you know? I don't think they're right about ecological arguments. Scientifically, they're not, but are they coming from a place that makes a bit of sense from what their moral foundation is? Oftentimes, yeah. They're like, "I'm trying to keep my family in this area and this is the main job that's here." There's an argument to be made for that. I don't think that trumps the ecological argument but it's their argument.

Alright, let's put it out there. And let's have the two ideas fight in the same place and see which idea wins. Isn't that what we're supposed to do? That's what most media doesn't do. That's what I try to do with the *Biomass Monitor*.

KOLLIBRI: Has much changed since Trump came in?

JOSH: No. Obama was just as much in favor of it. So this is something that you can hold both parties equally culpable for if you don't like it. Obama got a pass for a lot of anti-environmental stuff. And then the same stuff they let fly under Obama, people are all of sudden concerned about because Trump. Well that's disingenuous. I mean, I'm not a fan of any of those people or any of those parties.

KOLLIBRI: No, me neither. Taking this a little deeper, I think that the whole discussion around renewables is always about, "Oh, what are these new things we can do, new things we can add, to keep powering our way of life," when it's like, well, wait a minute: anytime you're adding something, you're adding destruction. Somewhere, you're taking down a forest, you're plowing up a desert, you're digging up all these rare earth minerals for your batteries. They all come with a cost—an environmental cost. And should we be adding more ability or should we be figuring out, collectively, how to be using far, far less?

JOSH: That's not a popular platform.

KOLLIBRI: Right.

JOSH: But that's the discussion that needs to be had and it's not being had. And how does that discussion get spurred? I think that's going to come from environmental groups because those are the ones that people turn to. And people are asking, "What are they saying about that?" But they're saying jackshit about it because you can't fundraise around that. You can fundraise around, "Hey we're going to lug in solar panels and you can stream Netflix." You can't be like, "Hey, what do you want to *not* do, and to change in your life?" That's not appealing.

That being sad, those who are advocating for simpler lifestyles, they need to do a better job. We need to do a better job of pointing out the pros of that life. Not just the sacrifices. You may feel happier because you are living a slowed-down life. You may have closer connections with members of your community.

If your goal is to increase your status, economically or otherwise, using less doesn't help that. There are studies that show your higher level of consumption increases your popularity with members of your own sex or the other sex. So that's a real thing we need to look at. So there's a lot reasons for people not to want to do that.

So what is the argument in favor of that? The satisfaction of being able to grow your own food or get your own food locally. There's a lot to be said for that. I mean, it's pretty awesome. It's not like, "Oh, don't go in your car!" No, it's "Ride your bike because you calves will get stronger."

KOLLIBRI: Because biking is sexier than driving.

JOSH: Right? Your ass will look better. However you can frame it. Not just in sexual terms, but that's part of it. That really is.

But [as for] having those discussions, the environmental groups are not doing that because they get money. I played the foundation game. You have to do what the foundations want. Where does the money for the foundation come from? Oh, the giant corporations. The Pew Charitable Trust that funds so many of the environmental groups, their money comes from Sun Oil, from Sunoco. I mean, I'm sure some of the children of the Sun Oil family are well-meaning, but do you really think they're people who want to revamp the entire economic system? No, that doesn't even exist in their minds. No one who grew up and lived that way way, for the most part, could think that, so that's what ends up happening. And then you're another environmental group and you're competing for that grant money and that attention. Good luck!

That's why I write. Let me put information out there. I don't know what else to do.

KOLLIBRI: I know what you mean. Have you heard of John Michael Greer? Formerly known as the Archdruid? He ran the Archdruid Report. It started as a peak oil thing before going into other topics. He shut down that website after ten years and now he has a site called ecosophia.net. He was a really interesting person because he really was an archdruid [and] he knew the science. He knew the politics and he knew the druid stuff. It was a really nice mix. But anyway, one thing he talked about in terms of activism—and what you're talking about—is: what does the message need to be? He said that activists have become afraid of painting the big picture; of saying, "Let's think big." Let's not start with, "Oh, here's what I *can't* say because of what I *think* people are going to say." He says we need to push that aside. What's our big fantasy wish-list utopia? Let's put *that* out there. Start *there*. Not "Oh, how can I keep this small?" but "How can I make this big?"

JOSH: I think a lot of foundation funding does dictate that. They want your mission to be tied into these small deliverables, so yeah, I think you're right.

KOLLIBRI: But as individuals we do this too. Not even looking at the corporate funding, the grant funding. Because we're worried. Because everything has become this big political game. "How will this come across?"

JOSH: There are repercussions. When I was working on the biomass issue in Vermont I was literally the only one speaking up. I was just saying stuff. I was trying to balance because it was all like, "Biomass is good," and I was like, "Well there are some other sides to this." At the time, I was writing op-eds. I was contacting the journalists. I was putting out stuff. I would talk to folks in the environmental groups and they all secretly agreed with me behind the scenes but almost no one would say anything publicly and I was basically blacklisted. Everything has since been proven and it's pretty basic science. I'm just saying, "Here is the EPA emissions stuff." I'm just saying that it's comparable in many ways to coal. "Do what you want with this. If you still thing it's good, I'm not going to stop you." But no one was responding to this point. You get blacklisted for that. It's not the party line and it's dangerous and some Democrats supported it, and most of the environmental movement is a vacuum for the Democratic party.

KOLLIBRI: But yet the discussion *has* to be expanded beyond the narrow confines it's in right now because the solutions for survival on the planet —not just for us but for other creatures here—are going to require radical changes. So the discussion *has* to be expanded. I mean, we *have* to push it out further.

JOSH: Yep, but that's the thing. How do we address the fact that by doing so, you do get marginalized? That's real. We can't act like that that doesn't happen because everyone who tries that sees it happening to them. So how do we address that?

I had a friend who was out here visiting and he's very well-versed in everything from permaculture to peak oil, across the board. He's just the most knowledgeable person I've ever met. And every time I would ask him this question he would dodge it. He would answer everything else super in depth. And I was like, okay, but those of us who are speaking out against this are facing repercussions. How do you explain to someone to keep doing that? You feel good about yourself but then you remove all your economic opportunities and people distance themselves from you? I think this has to be addressed.

I've had this happen and I've seen it happen to people who speak out. That's easy when those people are dicks, because people can say, "Oh, he's a dick, we're distancing [ourselves]." But I tried really hard and I'm not really a dick. Maybe a little but I'm not a dick to people and I'm not mean to them. I just put out contrary information and they didn't like it. But still a lot of people were like, "I don't want to be affiliated with what you're saying." So they all talked to me behind the scenes. This was back when I was activist. Some of them still don't want anything to do with it.

There's a danger to being those first individuals. And I'm really not trying to discourage this. I'm just pointing out that that's one of the biggest obstacles. How does an individual acknowledge those risk and push those risks and minimize those risks, and how do you support those who are taking those risks? I don't know.

KOLLIBRI: Individually, one could start with a question of, "Well, what's the worse that can happen to me from being socially marginalized?" Everyone's going to have different answers to that.

JOSH: That's a very good question. No, that's it. "What is the worse that can happen?" Totally...

KOLLIBRI: I guess I would say that the conversational space has to be held open. It just *has* to be and so we just need to support the people who are willing to take it to the fringes and to the margins. And personally, I don't have an issue with that anymore. I'm nearly 50 and for *me*, there have been no real consequences

JOSH: That's awesome. A lot of it is just the fear I suppose.

KOLLIBRI: And it's because I don't have the interest in the social status or things in that way, for whatever reason. I can't take credit for that, or say, "Gosh, that makes me cool," and I can't give you a step by step. Maybe that was just a constitution I was born with.

But I think that there's other ways of finding affiliation, or company, in a wider way. Currently, there's people all over the world who are trying to widen the conversations, or who *are* widening them, outside of the United States. There's people all around the world who are seeing these things more clearly. I mean, we are heavily propagandized here. We are one of the most heavily propagandized places in the world.

JOSH: Yeah. I mean, the information is here for us, but we're bombarded with all the other crap that you don't even see it.

KOLLIBRI: Right. It's like Chomsky talked about where you keep the area that you can discuss very narrow but then really actively discuss within it. That's what happens in this country. So first of all, that's just here. and in the global context, it's not that closed. It's not that narrow. Look at the indigenous people who are willing to go out there and put their lives on the line to save the forests they're living in. That's a much bigger worry than being socially marginalized in a place that's as wacky as this.

And secondly, I think you can also look at things in historical terms. Since destructive human behavior really started—the industrial revolution or agriculture—the whole time there's been that mainstream going on, there's always been a sidestream—continuous and uninterrupted—of people who have been witnessing it from the margins, from the outside and have been like, "I'm calling it. Here's what's happening. Here's the dangers. Here's how it's going to go down if we keep doing it." They've always been here. So in that case, you can be like, "Oh, *we've* always been here." I would

---

* *The exact quote: "The smart way to keep people passive and obedient is to strictly limit the spectrum of acceptable opinion, but allow very lively debate within that spectrum" [The Common Good, 2002].*

say that is one way in which people can reassure themselves. That's one way in which the marginalized voice is less alone.

JOSH: Right. I think that's a good point.

KOLLIBRI: Solidarity in time, and solidarity around the world. And fuck what people think in this selfish crazy place that we live in. You know what I mean?

JOSH: Yeah. And here's the other thing. It doesn't have to be everyone. It only has to be a certain percentage. Probably more than it is currently. There's probably going to be a lot of people who will never step out like that, but not all of them have to. That can shift the whole thing.

So in my journalism—of course I have my own bias. I have my own personal opinions—I try very hard to get not just "both" sides, but many side of the story. Of course just the way I'm framing stuff, I have a pro-environmental bias, but I get the other side's argument, and, I get how a lot of times, the environmental movement is full of shit. Not in the way that most people would say; in the way that's it's disingenuous and it's not actually pushing for what it says its stands for.

I think most activism is failing and I think its' unwilling to look at how its failing. So toward the end of my activist career I spent a lot of time just being like, "Here's how we're losing," and them people were like, "Why don't you support our cause?" and I was like—

KOLLIBRI:—"that's not what I said." [*laughs*]

JOSH: [*laughs*] "Wrong!" So I kind of got tired of that. So I was like, "Maybe I'll just put it out this way, and this way I can actually frame the arguments that activists are making in a structure that people can take a look at." I can't write anything without activists. That's where I find all the information out from. But I realize my role is no longer that.

KOLLIBRI: I think that when it comes to the environmental movement—for want of a better word—I think that it is largely failing. If you look at the last 50 years, you know—at the beginning of which we got the Clean Air Act, the Clean Water Act, the Endangered Species Act, the EPA, all put in a place by a Republican because of massive pressure—

JOSH:—he had to—

KOLLIBRI: Yeah, he had too—and that basically it's all been downhill since. It's all been watered down since then.

JOSH: I would tend to agree with the that. There hasn't been any big comprehensive changes.

KOLLIBRI: I mean there's less CFCs and so the ozone is not in as bad of shape. So there's *some* things. But by and large the direction is a declining environment.

JOSH: Yeah, if you look at one argument: "Well, there's less logging of old growth forests than ever before." Yeah, because there's only a fraction of it left. That's not really a fair measurement. We have ten percent of it left. It wouldn't even be possible to log at that level anymore. You have to look at it that way, at this point in time, with ecosystems in collapse—

KOLLIBRI:—species extinction—

JOSH:—even little things are a big deal at this point. So who knows. I'm not sure.

KOLLIBRI: All of that's getting worse. Basically environmental degradation has been getting worse. But there's this *idea* that is hasn't, or isn't, or—?

JOSH: There are little things that have improved here or there. But I think if you look altogether at biodiversity, land base, species—there's no question that it's been declining. I don't think there's anyone who debates that.

KOLLIBRI: And a climate going crazy at this point.

JOSH: Yeah.

KOLLIBRI: No one can argue against it but no one is acting like it's going on. At least here [in the US].

JOSH: No, it's true. But Americans are very comfortable. And [*sighs*] I mean I get it. You have kids and you want to take care of your kids and like, I get it.

KOLLIBRI: I think American are getting less comfortable, though. I mean, statistically, wage earners are making less than in in the early '70s.

JOSH: But compared to the rest of the world, a poor person here lives like a king.

KOLLIBRI: True. But look at the number of Americans now who are poor, officially poor; the number of people on public assistance—even with them trying to cut down people who can be on it—which was a big Clinton thing: to cut people off. Even with that, we have record high numbers. And now, things like infant mortality are all going up. The average lifespan is now declining. This is the direction that things are starting to go now, economically.

There's a class of people who went to the right schools and have good jobs and are comfortable, whose values and lives are represented—overrepresented—in the media, both the news media and the entertainment media, so there's this perception of comfort there. But that's a small class of people because the larger percentage of people are becoming less comfortable at this point. I mean, if you have to work *x* many jobs—

JOSH:—cost of living is going up, wages are going down—

KOLLIBRI:—How many states is it where a forty hour job at minimum wage will actually get you a two bedroom apartment? What is it, like six states?

JOSH: But we have way more cool stuff. That's real. People have their stuff.

KOLLIBRI: Apple has been exempted from the tariffs on imported Chinese goods.

JOSH: Of course. So that's all part of it. And I'm not pretending I'm outside of that. Certainly not right now. Definitely when I was younger, I did live more on the fringes, but I also felt more ostracized. I try to live, probably, simpler than most Americans, but compared to people around the world, I live like royalty.

KOLLIBRI: Globally, the number of people in the US who are classified as poor is two percent. Everyone else is middle class. By global standards we're middle class. Me living out of a pickup truck is middle class. Because that's amazing to have that thing! [*laughs*]

JOSH: [*laughs*] Right. But insofar as the trend is downward, in many ways, it's hard to deny that. And of course, inequality. It's not necessarily about poverty level. It's when you compare yourself to other people. So like Haiti is a very poor county, but people [are] happier because most people are equally poor. If you're in an area where there's a fair amount of rich people and a lot of poor people, that makes people unhappy. Inequality does.

KOLLIBRI: It's crazy inequality here.

JOSH: Yeah we're doing it. People are seeing it far more. I don't know what to do about that. I don't know what to do about any of it, really.

KOLLIBRI: Can you think of a big picture fantasy utopia you'd like to see?

JOSH: [*pause*] All I know is we have to get closer to the land. While I am a fan of the wilderness—I go out to the wilderness a lot and it's what keeps me sane—but I go out there and I'm like, "I shouldn't have to go nine miles by myself where I can die just to be in nature." Once a upon a time, a community of human beings was also in nature. So it would nice to have—and I like to get the hell away from people, frankly, too—but I would like to see more situations where we can combine human interaction with the natural world. And then we have the best of all worlds. Because I don't think big city living is healthy. I don't think it is at all. And I understand why people live here and I moved here for some of those same reasons.

So finding that nice balance between the natural world and human communities. Modeling aspects of hunter/gatherer time. Obviously we literally can not go back to hunter/gatherer right now with population size and the limited amount of ecosystem availability. People are just like, "Oh, I'll hunt deer." Well yeah, *you* get to, but if everyone in Denver decided to do that, there's no more deer. So that's not a wide scale solution for society.

But yeah, the land. The land is what we need. Beyond that, we don't even know how to do it. People try to start up intentional communities but because we haven't dealt with our *human* difficulties, they fall apart. The ones that work best are actually religious cults. I'm not in favor of that, but there's something to be said for that. So I'm basically saying a cult is the answer. [*laughs*] Cults. It always comes back to cults. But like a non-culty cult. Something to believe in. A unified belief system is what keeps people together.

These things fall apart because people don't deal with our human stuff and I think that the activist community is really bad at looking at those things. "Oh well, we're working for a pure holy cause." And it's like yeah— but you're just such an asshole. We all are, to a certain degree, and until we look at that, that's a whole different thing. In some of my discussions around environmental stuff, I've been like, "And we should meditate and go to therapy," and they're like, "What does that have to do with saving the forest?" and I'm like—

KOLLIBRI:—everything!

JOSH: A lot to do with it! Kind of a lot. Strangely. That's all I know.

KOLLIBRI: Right. Is there anything else you want to say while this [recorder] is running?

JOSH: Join my cult! [*laughs*]

September 2018

Cross-Country Sprint, Part 2

ILLINOIS

August 27th

I found myself on the I-65 in downstate Illinois, on my way to visit a friend in Urbana-Champaign. The land was some of the flattest I've ever seen. No hills as far as the eye could see. The horizon was a totally straight line.

Seemingly every acre of scenery was filled with crops, mostly corn and soy. Trees were few in number and tended to be clumped around farmhouses or along watercourses. At one time in the not-so-distant past, there were more trees, but as smaller family-owned parcels were absorbed into larger, corporate-run operations, they were razed for the "fence-to-fence" style of agriculture that now dominates.

And dominates it does in Illinois. People in cities tend to think of urban sprawl as a major environmental problem, but agricultural sprawl is far, far worse. For one thing, it's footprint is larger by orders of magnitude. For another, most of it is subjected to intense chemical treatments with substances that aren't even legal in cities.

I was amazed by the such vast areas of botanical conformity and pulled off at a random exit for a closer look.

As it happened, there was a cemetery with markers going back over a century. One stone said "Bushman" and I couldn't help but chuckle. "Bushmen" are exactly the kind of people wiped out by agriculture, so didn't it make sense to find a grave for them here?

Fake flowers decorated some of the plots. I remember when only real flowers were offered. How long do the fake ones stay out? The real ones had a distinct duration measured in days, and then a caretaker would clear them away. Plastic doesn't last forever either and eventually fades or starts disintegrating. At some point, someone has to make the call that they're trash and throw them away. Then what? They're buried in a dump, incinerated in a furnace of float out to sea. In other words, they pollute the land, the air or the water. Yet again, an "improvement" that is no such thing. A few plots were also marked by perennial plants. Those are from another time again, before my memory. It's probably against the rules now.

Beyond the cemetery, some distance away from me, stood a group of shiny metal grain silos, round with conical tops, squat in shape but rather tall; I would guess at least 40 feet high. Each one had a crane-like structure built beside it for filling it up. In rural areas like this, grain silos are the tallest things for miles and miles around. The skinnier ones can look like skyscrapers in the distance.

From time to time grain elevators explode. The dust is highly flammable and can be ignited by heat from machinery or even just static electricity. One spark can totally flatten one of these suckers. According to Purdue University[2] there are an average of 8.4 grain dust explosions per year. 2018 saw a dozen, with a total of four injuries and one fatality. In the history of such events, 1977 was a particularly bad year, with the Westwego grain elevator explosion of Dec. 22 in Louisiana causing 36 deaths and the Galveston grain elevator explosion in Texas just five days later responsible for another 20.[3] So here by this cemetery were placed these potential bombs. I'm not sure what kind of juxtaposition that is; the usual adjectives —poignant, ironic, etc.—don't quite fit, but it's... something.

Across the interstate from the cemetery was another kind of threat: a field of Soybeans. The crop looked ready to harvest soon; the pods were mature and drying out, hanging off the stems like sausage links. Given that over 90% of the soybeans grown in the United States are genetically-modified organisms (GMO), that's most likely what I was looking at.

Despite claims to the contrary, neither Soybeans nor any other GMO crops have been engineered explicitly to produce higher yields. Soy has been genetically modified to be "Round-up Ready," which is to say, resistant to the herbicide glyphosate, the active ingredient in Round-up. Both GMO soybeans and glyphosate are manufactured and sold by the notorious Monsanto Corporation. As more Round-up Ready crops have been planted, more Round-up has been applied. Why? Because weeds have been modifying themselves and developing their own natural resistance to glyphosate. This has necessitated higher applications of Round-Up and the reintroduction of other pesticides like 2,4-D.[4] 2,4-D is a nasty one; it's been implicated in birth defects, cancers, and thyroid issues in humans and is toxic to water-dwelling creatures.[5]

Monsanto has long denied that glyphosate is harmful to humans but in 2018, the tide started to turn against them. They lost a case in California in which a jury found that Round-up was responsible for a man's cancer and the company was ordered to pay millions in damages. Other such cases have followed. Additionally, Monsanto continues to be exposed for their interference in the science, their misrepresentation of facts and their meddling in the regulatory process to stay off the hook for culpability.

Regardless of what glyphosate does to humans, its environmental effects are beyond dispute. No one denies that it kills plants. That's what it's for, after all. But when you kill plants, you affect animals. Birds, mammals, reptiles, insects, and spiders are among the creatures who lose habitat and food sources when vegetation is destroyed. Herbicides also don't distinguish between native and non-native plants (nor indeed do many animals, who come to depend on exotic species when the indigenous ones decline).

Populations of the iconic Monarch butterfly have shrunk dramatically since the turn of the millennium in part because their primary food source, Milkweed (genus *Asclepias*), is being wiped out by the increased use of glyphosate.

The toxicity of glyphosate is compounded in water. Fish, amphibians and insects exposed to the chemical in riparian areas are harmed in a variety of ways, including genetic damage and immune system disruption. As these species suffer, so do others in the food-web, such as the birds and mammals who eat them.

Monsanto is scamming us and killing the planet for the sake of their own profit. Given the scale of the damage, we are justified to ask: Should any means be left off the table when it comes to stopping them?

Gazing out over a monocropped field, GMO or not, is surreal. The plants are not exactly identical, but close enough. So your field of vision is taken up by the same object repeated thousands—tens of thousands, hundreds of thousands—of times, for miles, maybe to the horizon. In wild settings, plants don't grow like that. Monocrops don't exist in nature. It's not a healthy way for plants to grow, so they don't.

What is it like to be one of these plants, packed in such close quarters with so many of your siblings? ("Brothers and sisters" is not accurate because the flowers on every Soybean plant contain both male and female sex organs—an arrangement described by botanists, by the way, as "perfect"—so every plant is neither/both.)

What is it like when no one but your own kind is present? And when the other plants who try to grow are poisoned to death? Are you grateful that the "competition" is gone or would you prefer some company, perhaps some cooperation? Maybe your senses aren't as acute as those of your ancestors due to your modification, and so these are all moot questions. Is life in a monocropped field a lonely, numbing daze all day long? Or is the pain of injury and disconnection sharper?

Last but not least, what does it do to humans when we eat food produced this way? Or to the confined animals who have nothing else and who are in turn consumed by humans? What we have here is a closed system of domestication, with genetic modification and chemicals severing some of the last ties between living creatures and their environment. For some crops, the very soil is sterilized so that not even microorganisms are part of the picture anymore.

But there's one element that humans cannot control, much as we might like to, and that's *the* elements. You can't poison away a flood or a heatwave or a changing climate. As I write this (mid-June of 2019), floodwaters have still not completely receded in the Midwest and a large percentage of this year's crops have not been planted and soon it will be too late. At what point does this massive system break down far enough that people don't have enough to eat? Famines have been regular occurrences all over the world, throughout the entire history of agriculture, and the amount of time we've gone without one here in the US is quite remarkable, and even unprecedented. That is to say, we're due.

CHICAGO

August 30th

As with NYC, I visited Chicago for just one reason: to meet someone in person who I only knew online. In the Windy City it was Paul Street, a writer whose work appears regularly on Counterpunch and Truthdig. Paul is a little more than a decade older than me, so he remembers the 1960s and early '70s, a period that holds strong interest for me.

I agree with historian John McMillan, who posits that the socio-political phenomena known as "the Sixties" started in the mid-50's with the Civil Rights movement and lasted until the end of the Vietnam War in 1975— which is also when John Lennon and Miles Davis, both emblematic of the era in different ways, retired from public life. McMillan calls this period the "Long Sixties."

I was born in 1969, so I missed all of it. I have just a few scattered memories from the first half of the '70s: a CBS news broadcast with Dan Rather talking about troops coming home from Vietnam; a line at a gas station in 1974 during the OPEC embargo, where my father, an economics professor, explained it; seeing a long-haired man in the grocery store, asking my mother if he was a hippie, and being downcast at her answer, "No, the hippies are all gone now." As my childhood passed, I had a vague sense that something special was fading away and when the '80s started, it felt like entering a dark tunnel (one that we still haven't exited).

So I appreciated conversing with Paul, who not only has first-hand memories older than mine, but has also taught history. That is, his leftist perspectives were informed both by academia and his own experiences. He was also receptive to different outlooks; just because I was younger and didn't have letters after my name didn't disqualify me from having something to say that could interest him. It strikes me that this is one of the most important things you can do as you age: stay open!

After lunch at a Mexican place, he took me on an activist tour of the Loop, the central area of Chicago named after the rectangle formed by the converging lines of the city's elevated train tracks. That month happened to be the 50th Anniversary of the notorious protests against the Democratic Convention there, and our first stop was the Hilton, site of both lively mobilizations and brutal police tactics during that event. Across the street in Grant Park was the horse statue made famous by a Life Magazine photograph of it swarmed by protesters.

We walked north along the waterfront park. Here was where Occupy tried to set up an encampment and over a hundred were arrested. There was the spot where anti-war activists rallied against the invasion of Iraq. We also saw cultural attractions including the lion statues in front of the Art Institute of Chicago; the four-sided mosaic, "Four Seasons," by Marc Chagall; and the post-modern architectural confection that is the Harold Washington Library Center.

Near the end of the tour, he took me by the Chicago Board of Trade and the Federal Reserve Bank of Chicago, which are both at the intersection of Jackson and LaSalle. That's where Occupy Chicago set itself up with a 24-hour presence, though without the encampment that characterized Occupy in most other cities..

As of this writing (July 2019), 2011's Occupy Movement was the last major outburst of protest activity in the United States (the brief flashes of anti-Trump "resistance" failing to rise to the same level). It's contribution to the popular lexicon of the phrase, "the 99% vs. the 1%," was definitely key. Whether or not those are the exact numbers is unimportant; what was brought out into the open was the fact that an oligarchy runs the US, something that many people across the political spectrum already knew or suspected, but which hadn't been proclaimed aloud so succinctly. Certainly, the corporate media has always avoided acknowledging this reality and has smacked down those who try to expose it. But Occupy let the cat out of the bag, and the cat hasn't gone back in again (as no cat would, given the choice).

The popularity of the Bernie Sanders campaign in 2016 drew in part—maybe in large part—on the spirit roused by Occupy. But the Democratic Party has been "the graveyard of progressive social movements" (as is so often said) since at least the 1890s and an electoral campaign is not a popular movement, no matter how popular it is, so the energy funneled into Bernie's "revolution" was ultimately wasted. Well illustrated, however, was the ruthless way the system chews up idealism and spits it out.

I had personally hoped that Trump's ascendancy would revive activism, especially the anti-war movement, but that didn't happen. Instead, a large percentage of liberals and the left took up the cause of Russiagate, a conspiratorial narrative that aimed to pin the blame for Trump's win on "Russian meddling" or "collusion" between the real estate mogul and Putin. Ignored were far bigger factors in the election: Hillary's profound unpopularity, her inattention to bread-and-butter issues, her lack of campaigning in the key "swing" states that lost her the electoral college, the billions of dollars in free press gifted to Trump by the corporate media, and, last but certainly not least, the widespread disenfranchisement of typically Democratic voters (especially Blacks) by Republicans at the state level across the nation, but especially in the swing states. This last factor—which was enough to swing the election all by itself—has still not received the attention it deserves, not even from Democrats, who were most hurt by it. (For the details of this particular travesty, the work of investigative journalist Greg Palast is essential.)

Considering the multiple interconnected crises we are facing—economically, politically, and most importantly, environmentally—we need mass movements now more than ever. So where are they?

Flash back to November/December of 1999, when the anti-WTO protests broke out in Seattle. It felt like something big was happening, and it was. That event was the US-based manifestation of a worldwide phenomena, the Anti-Globalization Movement, whose rallying cry was *otro mundo es possible*—"another world is possible." Though that sentiment had been expressed in the US before—whether earnestly, by the Students for a Democratic Society in their 1962 Port Huron Statement; with revolutionary intent by the Black Panthers in their 1966 Manifesto; or ludicrously, by Abbie Hoffman and friends with their nomination of a pig for president in front of the Chicago Hilton in 1968—one of the last times it was taken seriously in the US was in Seattle, 1999, and in the score of months that followed.

I was working a temp job in an office in Stamford, Connecticut, when news of Seattle broke. My reaction was immediate: "It's here! It's here!" by which I meant the Sixties-style uprising I'd been waiting for since I was a teenager. I knew I had to get in on this as soon as I could.

Less than six months later I had broken up with my lover, jumped off the track I was on, and had dived headfirst into activism. The Ralph Nader presidential campaign served as my introduction to an array of people and scenes, including Greens, socialists, anarchists, animal rights activists, and polyamorous pagans. It felt like everyone was feeling the positive jolt of Seattle. Change was in the air.

I started posting stories and photos to Indymedia.org, a self-publishing activist news website set up by anarcho-techies in Seattle for the protests. Their original intention had only been to set up something for the WTO event, but the on-the-ground, from-the-people, non-corporate coverage trumpeted on the site proved so popular that they kept it running after that show was over.

For those too young to remember—and as a reminder to those who are old enough but missed this part of the story—in 1999, it was not easy to get your words on the internet. This was before social media and blogs. Sure, there were discussion forums, and threaded message boards, but these formats didn't lend themselves to news reporting, certainly not with multimedia elements.

That was the Indymedia innovation: a platform, free-of-charge, for individual expression that didn't pass through a filter. You could go to the rally downtown, stir up some trouble, and then come home and write about it. You could post your photos too, and, as the code was developed and the infrastructure improved, upload audio and video. After clicking the "PUBLISH" button, your story would appear online within moments for everyone to see. Nobody approved it first. Or proofed it either, and this showed, but that too was in the spirit of it all. Your truth was your truth regardless of how you spelled it. Some of the Beat Movement's "editing is lying" ethos lived on here, but more overt was the rebellion against corporate slickness. The educated classes have always tried to impose their standards as the only way to be respectable or "be understood" or "get things done," but that's always been cover for the fact that they're pricks who don't want to share power. At its best, Indymedia freed people from that tyranny and set no voice above another. At the very least, it served as a check on corporate media, especially at the city level. If your local CBS affiliate reported that there were only "dozens" of protesters at an event, you and your comrades could set the record straight on your local Indymedia site with your photos showing hundreds. In Portland, where I was active with the local Indymedia in the first five years of the millennium, local corporate media felt enough pressure from us that they were forced to respond. When they start attacking you, you know you're on to something.

In the history of activism, such moments are rare: that is, when activists offer something people want—in this case, access to internet publishing—and enough friction is produced to make some heat. Indymedia thrived while it had this market cornered, but that didn't last long. By 2005, blogs and social media were muscling in on the space, and by appealing to peoples' ego, quickly gained success. The anti-war movement, too, faltered. After the Iraq invasion of March 2003, the big marches had ceased.

But the death knell for activism was the election of Barack Obama, and here, Paul Street has had a lot to say. He's been following Obama since his days as a local politician in Illinois and early on he spotted him as a neoliberal with no real commitment to progressive causes. For my part, I predicted in 2008 that Obama would turn out to be a "war-mongering Uncle Tom." That didn't go over too well in Portland, where the great black hope garnered over 75% of the vote.

The eight years that followed proved me right, and worse. All the liberals seemed to be assuming that everything was okay since their guy was in there and that they didn't have to pay attention to anything now. What did they miss?

- The continuation of *all* the wars W started
- Legalization of W's illegal power grabs
- Escalation of the drone program
- The sacking of Libya, then the most prosperous country in Africa
- Stationing military in nearly every other African nation
- Expansion of the surveillance state
- Attacks on whistle-blowers at the highest scale ever
- Increasing fossil fuel production to highest levels since the early '70s
- Refusing to hold Wall Street accountable for their crimes and culpability in the 2008 crash
- Gutting of the Voting Rights Act
- Supporting Neo-Nazis in Ukraine
- Deploying Special Operations to 134 countries (compared to 60 by Bush)
- Expanding the nuclear arsenal with a $1 *trillion* program for "modernization"
- Virtually ignoring Climate Change
- Last but totally not least, repealing the ban on using the US news media to distribute government propaganda

That's not a complete list, but it hits the highlights, or rather the low-lights. The only rays of hope were Occupy and Black Lives Matter and of the two, only one is still going. I will go so far as to say that Obama's two terms were the eight most wasted years in US history. While the wars raged and the planet was trashed, supposedly good people did nothing. What is the price that we will all pay for that inaction? For the children killed by US bombs and for the ecosystems sacrificed to fracking at Obama's behest, the price was the ultimate one. Countless crimes were committed that can never be taken back. Precious time was lost forever. Blood is on the hands of the Obama supporters.

Now an openly fascist person is in the White House, and most people seem like deer frozen in the headlights. Where is the action? The *real* resistance? The fight that's needed? My fear is that there won't be any.

On the big political subjects of our time—militarism, capitalism, ecocide—Paul Street is one of the smarter people around right now. He's no partisan and his knowledge of history and his own participation in the movements of his times inform his understanding of where we are and where we need to go. I highly recommend his work.

Back in front of Paul's apartment, I snapped a selfie of the two of us and took my leave. Altogether, we had hung out for only three hours. Back on the sluggish, rush-hour interstate heading north out of town, a particular sense of melancholy came over me that I didn't experience before or after on that trip. It was the bittersweet sensation that, if I lived in Chicago, Paul and I would be friends, but that I don't, and that our paths will probably never cross again. But this is a classic risk/benefit of traveling: that meetings will be so enjoyable that their partings bring a tear.

THE TWIN CITIES

September 5th-8th

The "Twin Cities" are Minneapolis and St. Paul, adjacent cities in Minnesota, along the Mississippi River. I lived in Minneapolis twice: in the early '90's and again at the turn of the millennium, so returning for a visit was a homecoming of sorts.

I met up with two of my best friends from high school. We had spent an untold amount of time together as teenagers, both in and out of school, but I didn't know what it would be like to see each other after twenty years or more. As it turned out, it was pretty much as if no time had passed at all. We just got right into talking without barely a flutter. In a few minutes, the flow was totally easy, like it had always been.

I knew that many things had happened during the intervening years—relationships, jobs, moves, scrapes with death (in their cases, not mine), having children (for one of them)—and that we weren't, in some way of saying it, the same people we were before. But none of that got in the way of connecting. Which makes me wonder: how meaningful is it to say that we weren't the same? Were we all, in actuality, fundamentally unchanged? Did the ease of our coming together demonstrate that the very idea of becoming a different person is a false one? Was I kidding myself when I looked back over my life and considered that there were old things I had shed and new things I had taken up? Or, if I *had* managed to make those accomplishments, were they only very minor by nature, just a scratching of the surface? Had I, in fact, never seen into myself with any depth at all?

I also met up with someone entirely new: Nick Pemberton, a Counterpunch writer and recent grad of Gustavus Adolphus College in St. Paul. I'd been following his work online and liked both his outlook and his writing style, the latter which I would describe as "breezy" (as in easy and open, not lightweight). I ascribe both to his youth. I have a heartfelt appreciation for the Millennials, their struggles, and their freshness. They are mocked too much by the stick-in-the-mud oldsters, including my generation, the Xers. After a snafu about where we were meeting (my fault), Nick and I got together at a coffee place in Northeast Minneapolis. I enjoyed his company and hearing the things he had to say. Coffee went well enough to continue at a brew pub, which then launched us into dinner with one of my high school friends and his wife. My friend has always been in a higher income bracket than me so he paid for my meal *and* for Nick's, which was appreciated since his choice was a trendy tapas place that was totally out of budget for both me and Nick. The meal was a *prix fixe* in five courses and he also ordered a flight of tequila to sample. A fun day had morphed into a lovely night and I slept well.

These kinds of experiences—coffee houses, brew pubs, trendy restaurants—had been out of my ordinary life for nearly twenty years at this point, since I left the East Coast. My spending had become much more frugal during my activist years because I was no longer willing to spend enough time working for other people to make that kind of money. This day was exceptional in that way.

I crossed the river for the second part of my stay, to see a friend from college. She lived in my old neighborhood in south Minneapolis so I revisited streets I had walked down literally hundreds of times before. There, I felt a sense of distance. Before turning a corner, I tried to guess what I would see, and more times than not I didn't know. This was largely due to all the change since last time I'd been there. The wave of gentrification hadn't started until after I left, but once it did, it had been busy. A few old standbys still existed—the Wedge Co-op, the Leaning Tower of Pizza, the Black Forest restaurant—but so many businesses, and in fact entire buildings, were gone or new.

Tragically, Muddy Waters was no longer in business. It was the first coffee shop I ever went into, back in the late '80s on a visit from college. This was in the pre-Starbucks era, when a cappuccino was hard to find but cost less than a buck and a half. By the mid-90's, coffee shops were springing up all over Minneapolis and I had seven to choose from within a mile of my apartment.

I almost lived in coffee shops in those days. You could smoke in them then, and nurse the same cup for hours if you wanted. Often I brought a notebook and wrote. Sometimes I read the paper or just stared out the window, too. But I always engaged in people-watching.

Coffee shops were places to meet people back then. Laptops didn't exist yet and internet was something you had at work or not all (unless you were an AOLer). I regularly struck up conversations with total strangers, and they regularly struck them up with me. Many I would only see that once and never again. Others became acquaintances or friends, and in one case, a lover. (But the story of Hugh—a fellow Nebraska guy who was sweeter than I deserved and who ended up killing himself, so whenever I hear his favorite song, "Dream a Little Dream of Me," I tear up a little—is for another time.)

As I said, this was pre-Starbucks, so every coffee shop was different. Decor, selection, service, clientele, price: these all varied, and one day I would be in the mood for one and the next day another. Sometimes I would pass a couple coffee shops to visit a third. I remember well-lit spaces with big windows and white walls; grungy ones with scuffed up counters and squeaky stools; comfy downstairs ones with heavy tables and overstuffed chairs. It was a golden age of its own kind, and I knew it would end, though not when or how.

As it turns out, the end was a decade away and the culprit was portable technology, but this, too, is a subject for another time.

Those days—the mid 90's—were the the tail end of a golden age not just for coffee shops, but for US cities in general, at least from a certain point of view. You could still find a cheap place in a cool old building in a central location, find an okay job pretty easily, and then spend your free time meeting cool people in cool hang-outs. Such a lifestyle was encouraging of creativity, be it writing, painting, music, dance or whatever. Sexual energy flowed through all of it, too, and in Minneapolis, the range of options was wide for all of the above.

These days were the end of a time that peaked in the '70s or '80s. That was when the conventional embrace of the suburban and rejection of the urban was at its height; when all the squares were out on the edges, leaving the city's heart for the square pegs. "*Stadtluft macht frei*," the Germans used to say, during the 11th and 12th century re-urbanization that followed Europe's Dark Ages: "City air makes you free." In the mid-90's, it still was, when I would order a cappuccino in Cafe Wyrd and check out who was to check out.

Gentrification swept it all away, but made a cartoon of its memory, trademarked it, and hung it on the wall in a reclaimed frame. "With a bird on it."

It started in the bigger cities first, then filtered down. So San Francisco and Boston got taken before Seattle and Minneapolis, and Portland after them.

Now cities aren't the places to go for cheap living. Increasingly, they are playgrounds for the wealthy and the professional. Where once there were artists, now there is a "creative class."

This is all part of a multi-decade trend that began with the great suburban build-out following WWII. This was the result of policy, not a free market. The feds began guaranteeing housing loans, and the terms of the real estate industry changed in a serious way. Where once you had to save up half the price of a house and had only five years to pay off the rest, now you only needed to come up with a 10-15% down payment and could settle up over twenty to thirty years. Also, these favorable terms were extended only to new housing out on the fringes, not older stock in the cities, especially not in the ghettos. Certain neighborhoods were circled in red ink on maps and no loans would be extended there; hence the term "red-lining." The population started to move out, at least the part of it whom banks would lend to, which was not everyone, with financial considerations being only one criteria. Skin color was high on the list, often at the top.

Then, in the late '50s construction on the interstate highway system began. Nearly all the old city centers were pierced and ringed with high-speed lanes, some trenched, some elevated (both styles with their own ugliness). Neighborhoods were carved up, homes and businesses demolished, communities broken. Real estate prices fell and crime rose.

The spaces that opened up offered both risks and rewards. Freedom flourished for a few, though again, the color of one's skin made a difference.

Throughout the '60s and '70s and into the '80s, cities were affordable. Artists, musicians, political activists and other fringey folks set up shop in old warehouses, rented rooms in subdivided mansions, and explored alternatives together. The cities of that time spawned counter-culture and fostered freedom of expression. The impact on the culture-at-large was noticeable in fashion, art, politics, and family life in ways both subtle and rev-olutionary. For demographics who were not straight or white or rich, there can help to conserve them,were neighborhoods one could call one's own.

Then the pendulum swung back. The normies started returning. The prices went up. Everything got tighter. Before too long, the rent doubled and then quadrupled on that cheap apartment, the artists' warehouse became a pricey tapas place, and the coffee shop a web design firm.

The wave of gentrification has been nationwide and it is in the current stage of a multi-decade trend that is not complete. Perhaps the blow can be softened in certain localities through community organizing—rent control, zoning adjustments, tenants' rights, etc.—but while the system remains in place, the overall direction will not change. Tinker as much as you want with that machine, but its primary function is to generate profit for the few not make prosperity for the most. That's capitalism.

Now where do we go to live cheap, meet up, have fun and create?

I don't know. But if I find someplace, I'm not telling.

June 2019

NOTES:

[1] The Great American Stations. "Rutland, VT (RUT)."
     http://www.greatamericanstations.com/stations/rutland-vt-rud/ [retrieved 7/8/19]
[2] "Grain dust explosions up, injuries and fatalities decline" Purdue University Agriculture News, February 21, 2019).
     https://www.purdue.edu/newsroom/releases/2019/Q1/grain-dust-explosions-up,-injuries-and-fatalities-decline.html [retrieved 7/8/19]
[3] Wikipedia. "Dust Explosion."
[4] Benbdrook, Charles M. "Impacts of genetically engineered crops on pesticide use in the U.S.—the first sixteen years." *Environmental Sciences Europe* 2012, 24:24.
[5] Sedbrook, Danielle. "2,4-D: The Most Dangerous Pesticide You've Never Heard Of" (National Resource Defense Council, March 15, 2016)
     https://www.nrdc.org/stories/24-d-most-dangerous-pesticide-youve-never-heard [retrieved 7/8/19]

Methane, Media and Multidimensionality— Lunch with Jennifer Hynes

Jennifer Hynes is a climate/extinction activist best known for her work explaining the dangers of Arctic methane release. We became acquainted on social media through these issues and met for lunch in Boulder, Colorado, in mid-September, 2019. We enjoyed a wide-ranging conversation that covered the environment, human consciousness, the role of media, and agriculture, all in the context of "the end times."

KOLLIBRI: So, what is your take about how things are going?

JENNIFER: I think we're on the way down. I think there's no way out. I think it's too fast and too big. I'm giving edutainment commentary during the ride. As much as anything, that's my function. I'm not really trying to push for geoengineering solutions. I support people who want to do that but I'm in the witness seat.

KOLLIBRI: Sometimes that's all you can do.

JENNIFER: Yeah. I'm most well-known for my studies in Arctic methane. I'm the author of the "Arctic Methane Monster's Rapid Rise" and the "Demise of the Arctic."

KOLLIBRI: It seems like methane has been a big subject this summer—

JENNIFER:—yes it has—

KOLLIBRI:—because we're seeing new things.

JENNIFER: Yep. Things are really heating up and now there are perpetual hots zones on the East Siberian Arctic shelf. Methane measurements are kind of off-the-hook. I just saw a post from Sam Carena [of the "Arctic News" blog] today: 3300 parts per billion up in the Arctic.

KOLLIBRI: And historical levels are much, much lower than that.

JENNIFER: It's heading up quickly. But it's hidden. People still think every-thing's normal—anyone who's not really watching. It's given people the ability to continue on about their lives without mass panic, which is prob-ably nice, but it's weird because there's a few of us who are woken up and we're kind of like [*makes siren sound*]... I get the question a lot, well, if there's no hope, why bother telling people about it? And I'm like, well, that's a strange question. Have you ever gotten that one, too?

KOLLIBRI: Well, yeah, and I do think it's silly because I think it's mean-ingful that we're on a runaway train.

JENNIFER: We're witnesses and anyone who's doing this is turned into kind of a beacon of truth because there's very little truth out there. We're part of the same truth-teller tribe. And once you're part of the truth-teller tribe, you meet people and it's easy. You've noticed that, right?

KOLLIBRI: Yes.

JENNIFER: It's like, instant, because you all know the bottom line. No-body's under the misapprehension that everything's okay. Even though it sort of looks okay.

KOLLIBRI: Right. "It's a lovely day here in Boulder," or whatever.

JENNIFER: It's a hot day for September, but it's not so hot that you'd sit up and stare but that's because people don't study the earth. And they should.

KOLLIBRI: I feel like, as creatures on this planet, we're picking up on what's going on on. There are messages that are being sent to our senses, right? Our senses are picking those up, but how our conscious brain is in-terpreting those—or ignoring them—that's the clincher. I feel personally that the levels of depression that you see in society, for example, are re-lated to that. Because we're receiving messages from the creatures of the earth that they're dying, they're suffering, that we're collectively doing that to them and that they're crying out for help. And those messages for help are reaching us. They're coming into our eyes, they're coming into our ears, they're touching our skin, etc. It's just that in this culture, we're abso-lutely not raised to be aware of such signals. At all. But so far, the climate has just become more pleasant for many people.

JENNIFER: Right?

KOLLIBRI: Because winters aren't as harsh.

JENNIFER: I was hearing the latest from Scotland and they're all like, "the weather's great." They've been having this heatwave in Scotland where it never goes above 75 degrees F ever.

KOLLIBRI: So since you know so much about methane, I'm interested in more of your take on the details of what's been happening this summer, but also the big picture of what a massive methane release could look like.

106

JENNIFER: Right. Well, not to review the obvious, but the only thing holding the methane in place that's in these vast, vast stores in the Arctic is the ice cage that's around it.

KOLLIBRI: The permafrost?

JENNIFER: Well it's in the permafrost too, but I'm talking about in the ocean, in the east Siberian Arctic shelf in the ocean itself. There are methane fountains. When you look at this methane, it's actually pure white —its not in permafrost—it's pure white, solid, frozen methane hydrates.

KOLLIBRI: Wow, okay.

JENNIFER: There's no dirt in it. They want to mine it: Russia, Japan, China, and they can, and they can get pure methane hydrate which will thaw as soon as it's above freezing—

KOLLIBRI: Freezing for water? Like the same temp?

JENNIFER: Yes. It thaws at the same temperature because it's methane hydrate. It's encased in a cage of water that's ice. But the difference between a frozen molecule of methane hydrate to an unfrozen gaseous one is 1 to 165.

KOLLIBRI: Wow. Like the volume, you mean?

JENNIFER: Yes. Have you heard of the methane blow-holes up in the Yamal Peninsula?

KOLLIBRI: That place name is not ringing a bell.

JENNIFER: Okay, this is a really good example. The Yamal Peninsula is near the Kara Sea and it sticks out into the Arctic. There's nobody up there except reindeer herders and their sole means of existence is herding. What has been happening the last several years is they've been discovering these massive craters—60 to 100 meters wide—and they have perfectly vertical walls and they're perfectly round and they've got exploded earth all around them and it's just like a blow out. So at first when they found them, it was a new phenomena. It's only been going on for like four years now.

KOLLIBRI: Okay, I've seen pictures of these.

JENNIFER: Right. The Siberian *Times* is a really good pictorial resource for things like this. So they found these things and it became evident what they were methane blow holes. What they would see is that little mountains start growing up where there were not little mountain before. Over the years it goes like that, and—

KOLLIBRI:—and they burst open—

JENNIFER:—and when they burst open, the reindeer herders have seen them go off. They've seen them go off in the sky at night. There's a huge flame in the sky, a huge explosion, a huge light from the burn, because it comes out and it ignites and [*makes boom! sound effect*] and then it's gone.

That's it going from 1 to 165. You get all of that happening. It's dramatic. That phenomena is happening all over the Arctic but especially in the Yamal Peninsula because that area seems to be perfect for that sort of activity.

KOLLIBRI: So, the fear is because methane as a greenhouse gas is x times as powerful—

JENNIFER:—when it first comes out, molecule for molecule—I'm sorry if I sound like a school teacher—

KOLLIBRI: No, no, that's what i want to hear.

JENNIFER: Okay. Not only is it bigger but it's also more potent and it degrades quickly. So when methane first comes out, it's about 125 times more potent, molecule for molecule, than carbon dioxide. So it doesn't take much methane to have a big effect. Not only is it more potent but it's a quicker burn. So when it comes out, those area are getting hotter. With carbon dioxide, there's a delay. You know that, right?

KOLLIBRI: Yeah.

JENNIFER: There's a bit of a dispute—some people say ten years, some people say twenty—there's a delay between the time it's emitted and the time that the heat is realized in the atmosphere. Not so with methane. Methane is volatile. So those areas in the Arctic, they're already hot now. They're having heatwaves up in Siberia, in June, of 95 degrees! That's just causing more feedbacks: this much more, you know, causes this much more warming which causes this much more...

KOLLIBRI: I first heard about feedbacks reading Guy McPherson's stuff. He explains quite a few of them very well.

JENNIFER: That's right. And they're all interlocked. One affects the other. And what's going on now is methane. That's being seen as the high methane reading, high temperature readings, and the abnormal sea ice. There's no normal arctic sea ice in the Siberian Sea anymore. The Arctic Sea that was the most firm in the north of Greenland, that's cleared out.

KOLLIBRI: I saw that that was happening. That's been in the last month or two.

JENNIFER: That's been the big news.

KOLLIBRI: So a lot methane release could lead to an abrupt temperature increase.

JENNIFER: Right.

[Break while waiter comes up, takes orders.]

JENNIFER: I mean, this is kind of classic, right? Here we are, eating the best food on earth, drinking fabulous alcohol, sitting in one of the most beautiful places, and we're talking about the end times.

KOLLIBRI: I know. [*laughs*]

JENNIFER: This could be a good past life memory.

KOLLIBRI: Nice! [*pause*] But how abrupt could a temperature increase be? Are we talking literally overnight or, like, weeks?

JENNIFER: Oh yeah. Weeks.

KOLLIBRI: Weeks?

JENNIFER: Weeks. It could be weeks. Maybe up to a year. But it would cause havoc and chaos. Like, the jet stream is already messed up—but the jet stream would get more messed up. It gets stuck and hurricanes can get slower. They can deposit more water. They can get bigger like this one that's currently just past Hong Kong [Typhoon Mangkhut].

KOLLIBRI: It was a super typhoon. That was number nine of all the tropical storms and hurricanes at that point.

JENNIFER: That's right—that was one of the nine. Florence was one of the nine.

KOLLIBRI: There were five in the Pacific and four in the Atlantic.

JENNIFER: Right. Which is crazy. Wouldn't you think something like that would be in the news?

KOLLIBRI: Right?

JENNIFER: So that's a strange thing.

KOLLIBRI: Let's talk about Stormy Daniels.

JENNIFER: If you wanted to get people fixated on something completely meaningless, I think Trump's brilliant. I mean, what a great way to focus the attention of the entire world—I mean, anyone who's reading US politics. We've got real problems and we're talking about this, to the exclusion of everything else.

KOLLIBRI: I know. It's extremely frustrating.

JENNIFER: It's like an addiction. It's like a hit. It's so crazy that it kind of gets you off.

KOLLIBRI: I think that "Trump derangement syndrome" is a real thing. And it frustrates me because I'm like, "Can we at least look at what he's doing to, say, the endangered species act? Because while all y'all are talking about Russia Russia Russia there's creatures here who are close to the edge of extinction and might go extinct while you're cheering for Robert Mueller or whatever.

JENNIFER: Exactly.

KOLLIBRI: People aren't even looking at the truly damaging thing that he's doing. It's a personality thing.

JENNIFER: It's really crazy. And what a time to live and to actually be aware of the play of existence going on before your very eyes.

KOLLIBRI: I was thinking about doing this road trip last year and I wasn't sure if I was going to do it or not. And I was thinking about going out to California and doing some work and my friend was like, "No, I think you should go on this road trip now. Maybe next year you won' be able to do it." And so he gave me some money.

JENNIFER: That's cool.

KOLLIBRI: Yeah, it was really nice of him.

JENNIFER: So this is kind of like your America discovery tour at the end times?

KOLLIBRI: Yeah, I've lived in different parts of it before, but now I'm looking at it in a very focused way.

JENNIFER: Do you see any changes in the areas you use to know? In the atmosphere or vibration?

KOLLIBRI: Well there's been this whole thing going on nationwide of the cities being gentrified. So what used to be the inexpensive area in downtown areas are too expensive to live in. That's all part of a larger trend, of where everybody abandoned the cities for the suburbs in the '50s and 60's, and started to come back in the 90's and 2000's. That's something I've seen.

But I think the biggest thing I've seen—and everyone I've asked about this has remarked about it too—has been the cultural changes that have occurred since this thing [*picks up phone*]—the smart phone—came along, which is only eleven years ago now, but it seems to have changed the culture significantly.

JENNIFER: Well think how informed you and I are. For me, it's because of Facebook that I have all this information.

KOLLIBRI: Well you can have Facebook and you can have the internet without the mobile device too. The mobile device has been it's own thing. Like the social media is one part of it, and the mobile device is the other part of it, and they work together really well, and yes, if your whole point is to stay informed, they can be a great tool. But I have a friend who calls these things personal propaganda devices and I think that that's also a good description of them.

So is there anyone who inspires you?

JENNIFER: Yeah, there are a few people who inspire me. Teachers that inspire me are ones that look at this and contend with it from a social perspective, from a spiritual perspective, from a processing perspective. So, it's one thing to share news, which is great. But it's another thing to say,

"This is all the news and it's pretty crazy and messed up, so what are we going to do with that?" One person, who's in Canada, is Louise LeBrun and she is kind of a teacher to the end times.

So I guess the people who attract me, who are focused on communication, and stuff like that, are what I would probably call—even if they don't call themselves teachers—are what I would call teachers for the end times. Anybody who contributes to the overall picture for me, and I can take that information, and they don't even have to be doing a take on it. Paul Beckwith is a good example. He can put together the science and say, "here it is," and you can retrofit it into your own consciousness.

We are all each others' teachers. Anybody who chooses to take an active and expansive role and integrated awareness, who appeals to you. There's a bunch of beacons out there. Anybody who's in this mode—activated—and who has a good head and understands the science and is a fully functioning, actualized, seeking, aware human being, will take that information and put it through their own awareness and put it out.

I think those of us who are activated, who are awake, are also consciousness-changers. Like, we as a group are participating in the awakening of human consciousnesses, and that may sound like a big thing to say, but it's not, because I've seen consciousness change. Like when the "Arctic Methane Monster Rapid's Rise" was published four years ago, it went viral. It was carried in the Arctic News, it was carried all over the world. I was in shock. I had no idea what was happening. But since then, I've been sort of woven in, in a multidimensional way, to this network of consciousness. There's a tribe. I can feel it on the psychic level, and we're all sort of aware of each other, more or less. We all work together and it's an amazing sort of thing to have this happening. So we are the harbingers of human awakened consciousness. And it's going to cause a huge reaction. I've been seeing more and more stuff being carried in the mainstream media, but usually it's just—

KOLLIBRI:—it's just a little piece tucked in there.

JENNIFER: Yeah. MSNBC was sorry that they were so bold the other day. They had Chris Hedges on!

KOLLIBRI: Unbelievable! [*laughs*]

JENNIFER: Yeah, well, he's not going to be on anymore. Of course he was classic Chris Hedges. Are you going to have Chris Hedges be any different than Chris Hedges? Chris Hedges is who he is. He is not going to pull punches just because he's on MSNBC. If anything, he might actually make it go harder. Well, he did not take any prisoners and I saw it live when it happened because I watch MSNBC during the day. I have it on in the background while I work from home. I'm sort of a new junkie on that level. And

I'm like, oh my god, Chris Hedges is on! You can tell his voice instantly. You watch a Chris Hedges video, you're never going to forget his voice, or his delivery or his content. So he was on and he was on with this young Asian telecaster, a minor MSNBC figure I'd never seen before, and she's trying to steer him over here because he's like, well, there's no hope. He does the whole thing and she just tried to shut him down. And that's such a fumy thing: her trying to shut him down. Give me a break. So I don't think they'll have him back. He's too much trouble.

KOLLIBRI: A lot of journalists have been pushed out of the mainstream media, like him. over the last few years. He's at RT [Russia Today] now with his own show. It's a great show. I enjoy Lee Camp's show on RT quite a bit too: "Redacted Tonight." I don't know if you watch that one?

JENNIFER: Oh yeah, Lee Camp. He's awesome.

KOLLIBRI: It's a great show.

JENNIFER: Radical.

KOLLIBRI: He's putting on a show in Boulder sometime soon.

JENNIFER: That'd be great! He retweeted one of my videos, "The Demise of the Arctic." which is two hours and forty minute long. Watch some of his climate stuff and you'll say, oh, okay, he knows. I've seen him react to climate stuff, specifically methane. He's very methane aware.

KOLLIBRI: So you talked about two different trends you've been watching. One of ecological disaster—whatever you'd like to call it—and one of human consciousness. Are they related in some way, for you?

JENNIFER: Yeah absolutely. I mean, at my base, I'm kind of a seeker. I'm like you, interested in finding out information and synthesizing it and understanding truly what's happening. What are the trends? Where are we going? What's happening? I have the requisite tools to make that kind of analysis, as do you. One just has to lead to the other. The climate crisis happens with human awareness, and humans—to different level of awareness—will wake up to this. Now it's actually my belief that *everybody's* woken up to this.

KOLLIBRI: On some level?

JENNIFER: Right. So if you want to talk about it on a purely psychological, multidimensional level, this is what I think.

KOLLIBRI: Everyone already knows what's going on.

JENNIFER: It's too big!

KOLLIBRI: I get that. I dig that.

JENNIFER: Everybody knows.

KOLLIBRI: But at the level of your everyday ego consciousness—

JENNIFER:—right—

KOLLIBRI:—the ego conspicuousness doesn't want anything to do with it.

JENNIFER: Right.

KOLLIBRI: And see, I think that in previous eras of human history, we weren't blocking the world out in that way. We were conscious of those things. We were conscious of the state of the world.

JENNIFER: Yeah.

KOLLIBRI: I think that's the result of agriculture and technology, and all that, that we've alienated ourselves from nature.

JENNIFER: Have our ever done any investigations into the civilizations that existed prior to the end of the last ice age?

KOLLIBRI: Just a little bit. it's fascinating stuff.

JENNIFER: *Really* fascinating. Like not even just a little bit. But it's also completely hidden. I love to study the hidden. Not conspiracy stuff. I'm really not into conspiracy stuff at all, but what I am into is hidden truth. Everybody has a different opinion about this, but I will go out on a limb: I believe we are multidimensional creatures.

KOLLIBRI: I don't think that's a controversial thing to say.

JENNIFER: And I believe that consciousness persists.

KOLLIBRI: Uh huh. I don't think that's controversial either. *I* don't. [*laughs*]

JENNIFER: Right. The thing is, in the doomer community—and I'm not saying "God," I'm not saying that or anything—I'm just saying in the doomer community, atheism is at large. Like, rampant, aggressive, atheism is at large.

KOLLIBRI: Doesn't that often times go hand in hand with people of a scientific education or academic background?

JENNIFER: It does. But now, a whole different kind of person is coming into this understanding, not just the purely scientific person, but the person who has room in their awareness for these realities. People who have done multidimensional studies, people have have meditated, people who are interested in the transformation of consciousness, people who are interested in various dimensions or various gradients of increasing light, people who are understanding that this is a moment: a unique unreplicated moment that we're living through and that it's ours to witness. So, people who have a multi-banded awareness, who understand that one dimension can just lead into another and you may not be aware of it because you're not tuned to that bandwidth on the dial.

KOLLIBRI: Not tuned in and I would also say enculturated *not* to be.

JENNIFER: Right. Vibrationally locked.

KOLLIBRI: Yeah.

JENNIFER: I think there are more multidimensional awarenesses coming into this collective awareness that we have. That's kind of what I'm thinking. We're building a collective awareness. I have 2000 friends on Facebook and people have different interacting sects of those. All in all, we are building some sort of collective awareness for those who are interested in it.

KOLLIBRI: Was it T.S. Eliot who said, "Where is the wisdom that's been lost to knowledge, and where is the knowledge that's been lost to information?" So information—that's entirely its own domain, wisdom is an entirely other different domain, knowledge is something else again, and now we're living in an age where it's been broken down smaller than information: to data. "Where is the information that is being lost to data?"

JENNIFER: Right. And wisdom can only be kept inside a human: Wisdom-keepers. So there are going to be those of us who are going to get more and more and more awake and we're going to be the wisdom keepers. And what we do with it, who knows? What a time to be alive!

KOLLIBRI: Yeah, I've known from an early age that I was going to be alive in an interesting time. I knew I wouldn't have to worry about saving for retirement.

JENNIFER: A lot of us have known this from an early age. A lot of us who were talking about this stuff. I can remember knowing this from an early age. Like a Cassandra complex that gets born inside of you. A lot of us had these seeds of awareness of the time that we're in.

[*Waiter breaks in for a moment. The conversation winds around to the subject of the Climate Change activist group, 350, which was started by Bill McKibben.*]

JENNIFER: Bill McKibben been changing lately, have you noticed that?

KOLLIBRI: No, I haven't noticed that. Has he gotten more radical?

JENNIFER: Yes. I noticed it. I went to—what's that show that's in Marin?—it's one of these eco shows. They have everybody there. No confirmed doomers, but a lot of solutions people. Bill McKibben was there and he spoke, and I have to say I was like, "Really?" because it was grim and he was like, "I don't know that we're going to get out of this, but we have to try." It was very very telling; Bill McKibben is changing.

KOLLIBRI: He would have to if he's paying attention.

JENNIFER: He would have to because his 350.org thing is fairly ridiculous. It's good that it exists, don't get me wrong. But the whole premise that we have to get *back to* 350 is not going to happen. Ever. Completely. Like, ever. You can do carbon sequestration out the wazoo and you will not be able to get this back. You could do it on the scale of whatever "mission" and it still [won't work]. So that's the whole thing. They're greenies. They're greenies to the core and for the most part they don't want to hear too much. They want to hear enough to propel their mission, but they don't want to hear too much. Is that kind of how you feel they are?

KOLLIBRI: Yeah, I see that. You're kind of describing the liberal intelligentsia in general, who just don't want to go all the way there.

JENNIFER: They just want to do heavy petting.

KOLLIBRI: [*Laughs uproariously*] That's a good way of putting it, yeah.

JENNIFER: [*Laughs loudly*]

KOLLIBRI: I mean, the liberal intelligentsia has always been like this. I don't know if you've ever read Emma Goldman?

JENNIFER: No.

KOLLIBRI: She's amazing. She's a famous anarchist from the early 20th century. You would probably really enjoy finding out more about her. She has an autobiography. She was from Russia, I believe, but she lived in the United States. She was an anarchist. She got arrested many times for being anti-war, for passing out information about birth control, for pushing for the woman's right to vote. All these kinds of things. She talks about the conflict between the liberals and the radicals then, in the early 20th century, and—same thing. Could be talking about now. It's the same people playing the same parts: those who want to tweak the system and those who see that the system itself is the issue, that's its not reformable. And 350 is basically a liberal organization. "How can we keep all this?"

JENNIFER: You can't.

KOLLIBRI: You can't. And it's not like this [the System] is actually working either. This isn't actually making anybody happy.

JENNIFER: No!

KOLLIBRI: The materialism, the division from nature—none of that is working. It's making everyone really unhappy and we're seeing lots of really bad things, including this culture being really destructive to other people and to other creatures all over the planet, with no real internal resistance to that. Because who speaks out against US imperialism?

JENNIFER: Exactly,

KOLLIBRI: It's a very short list.

JENNIFER: No one.

[*The conversation turns to Jennifer's upbringing in Laos, where her father worked for the US State Department. They lived in the jungle, without all the connections to the rest of the world that are common today.*]

JENNIFER: There was no TV. I listened to short wave radio in the morning. There was the BBC that came on in the morning. And I became a big time reader because there was nothing else to do. It was hot outside and there weren't that many people. We had a a great library because my mother was an English teacher and she brought all her books over. We just read and we played games outside and did Girl Scouts. Things like that. It was a very wholesome existence. It was kind of like an existence that a kid might have had in the 1940's or 1950's maybe, before TV really took off.

KOLLIBRI: TV had an immense affect on culture but almost no one who is alive now understands because almost no on who is alive now remembers what it was like before it.

JENNIFER: That's right. TV is a drug,

KOLLIBRI: It is.

JENNIFER: It's like, narcotic.

KOLLIBRI: From the beginning it was, and they just got more and more sophisticated with it.

JENNIFER: It's a propaganda machine that we allow into our house and allow into our lives. We allow this crock of propaganda because it gives us something that feels good.

KOLLIBRI: It's not just the propaganda of information, of the news, either. It's the propaganda of the culture: what you're supposed to believe, how you're supposed to be, what's supposed to be important to you, how you're supposed to view relationships, how you're supposed to view yourself. That's even more powerful I think than the misinformation that's received through the news. There was a writer on Counterpunch who I met in Chicago: Paul Street. He also writes for TruthDig. We talked a little bit about the social media and the phones. And it's his opinion that they are doing more harm than good.

JENNIFER: No doubt.

KOLLIBRI: Because as much as some individuals might get out of it, the powers-that-be—the corporatocracy or whatever—is getting *sooo* much more in terms of social control and in terms of wealth generation.

JENNIFER: Exactly. Where everybody is at and they can just data mine the hell out of everyone. I do it knowingly, knowing that side. I'm a data analytics forecasting person. I'm an IT person. So I already know what they can do with all this information. It's all in a database but I don't care. I'm completely open. It's almost like a social experiment on my part. See what will happen. Like, for example, my friends list, I don't keep it private. I keep it open. This is an experiment with *glasnost*—opennesses—how open can you be. But it's a vast social experiment, this whole thing that I'm kind of engaged in, in Facebook. You're engaging in it.

KOLLIBRI: I am. The whole thing's a big social experiment that was foisted on us without any real preparation or announcement or any forethought on the part of anyone who was involved in it, except in terms of making money.

JENNIFER: It just happened. Facebook was a big surprise. it just happened. All of a sudden there's this thing.

[Another break, as waiter brings more food, takes dishes away, etc.]

JENNIFER: *[compliments Brussels sprouts.]*

KOLLIBRI: They're one of the more challenging vegetables to grow.

JENNIFER: Are they?

KOLLIBRI: Yeah.

JENNIFER: Are you a vegetable grower?

KOLLIBRI: I was a farmer for ten years.

JENNIFER: Really?

KOLLIBRI: That's partly how I came to some of these things too. I was an organic farmer.

JENNIFER: So from a farming perspective, what's your takeaway? What are you noticing? Given this climate debacle, in terms of that.

KOLLIBRI: There's a huge disaster that's brewing that no one knows, almost no one is paying attention to, and which could happen in any particular season now. I would not be surprised—really—if any season, something suddenly happened. It's only a matter of time until something major happens in the United States. The wildfires in Oregon this year were taking out portions of the wheat crop in eastern Oregon.

JENNIFER: It's terrifying

KOLLIBRI: Yeah. Partly what we have to insulate ourselves with here in the United States is that an enormous amount of what we're growing is for animal feed, not person food. So a huge amount of it can go away and we can still eat, and just not have meat be as big a part of it. One fifth of the area of the lower 48 states is devoted to animal agriculture in one form or another. We don't need to eat that much, for health reasons, for anything else.

JENNIFER: So whats your vision? How do you think this thing is going to really play out? I mean, I've been called an alarmist for going to all these kinds of spaces in my mind. But how's it going to look from your point for view, from now until like 2030?

KOLLIBRI: I think that you have major agricultural disasters in the United States within that period, actually. I think that something happens in that period that wipes out a lot of crops. There's any number of ways that crops can get wiped out, but one thing I was noticing as a farmer—the last couple of years i was involved in it—was the chaotic, unpredictable nature of the weather. How we're getting more extremes, up and down, up and down. That was making it hard to know when to plant things and it was making it hard to know when to harvest things. Something like that is going to happen here for sure.

JENNIFER: So all of that means that crop yields are going to go down over time and maybe in a couple of years—

KOLLIBRI:—there will be some dramatic event—

JENNIFER:—then they'll recover and then—

KOLLIBRI:—things will never *fully* recover though.

JENNIFER: So how is that going to look from the geopolitical perspective in the United States?

KOLLIBRI: It's going to look bad here.

JENNIFER: And lets talk about neo-fascism in terms of crop failures an Climate Change. That's a potent thing right there.

KOLLIBRI: Cities are going to be really unpleasant places to be. I think that's where the worst of it is going to be. Any city has on hand—what?— ten days to two weeks worth of food, tops. Tops. Because of the whole "warehouse on wheels" that now exists in the United States, where you're not storing things. Everything is—

JENNIFER:—on demand.

KOLLIBRI: Right. On demand. You know, "just in time"—all that. So the rural areas will be easier places to live because you're not going to have this press of people all running out of food at the same time. I don't even know what's that s going to be like, in the cities. Because when people don't have enough food, they act differently.

JENNIFER: That's absolutely true.

KOLLIBRI: So it's hard to predict that.

JENNIFER: I grew up for some years in a fairly brutal place. I grew up in Ghana in the 1970's and there was a huge devaluation of the national cur- rency and there was famine and I can remember going into the stores— which were government-run—and there would be clean counters, people standing around, and nothing to buy. I can remember going to little gro- cery stores, and there was nothing to buy. Everything was very expensive —hyperinflation. So i can remember hyperinflation and empty shelves and that was an underdeveloped company going through very difficult growing pains. So I've seen what it looks like from that perspective. Of course, it didn't really strike me personally because I was in the *creme de la creme*. I lived an unreal life. If there was no food in the store, that didn't matter to me. We had a a commissary or whatever. We could send away to

Kenya for food, which we did. So local problems didn't really affect us at this very strange *creme de la creme* level, but I saw it nonetheless.

KOLLIBRI: The people of the United States have been living at that *creme de la creme* level for a couple generations now, but at some point that runs out. Then, lord only know what happens geopolitically, with a nation that has this many nuclear weapons, and how it can hold the rest of the world hostage that way. This nation has proven that it's willing to use them. We're the only nation that has. Which people here don't think about [but] people in the rest of the world know.

JENNIFER: And let's not forget, there's a raging sociopathic narcissist with his hand on the controls.

KOLLIBRI: Right.

JENNIFER: Which is terrifying, truth be told.

KOLLIBRI: I was terrified by either possibility for that reason, this last election.

JENNIFER: Yeah, either one was bad. Why couldn't we have had a normal candidate, or just not kill Bernie? I don't know.

KOLLIBRI: I don't think working with that system is any solution.

JENNIFER: No. What's going to happen is going to happen and its going down now fast.

September 2018

Life & Death on Route 395:
Water Wars, Baby Birds
& a Crashed Harley

I awoke the morning of my forty-sixth birthday on top of the Los Angeles Aqueduct. Not immediately on top of it—it was buried at a depth of at least 25 feet in this spot—but just a few steps from our parked van it was exposed by a service shaft: an open concrete box lidded with a thick grate. Standing over it, I could hear the roar of water being sucked southwards by the thirsty urban monster 100 miles away. Down the hill from my vantage point, the giant metal drinking straw of the aqueduct's pipe emerged from the ground, spanned the dry creek below, and thrust itself back into the opposite slope. Another service shaft stuck up out of the hilltop above it, just like the one in front of me.

The Romans made far more elegant structures for the same purpose, but I marveled at the feat of engineering laid before (and underneath and behind) me. I wondered about the degree of angle it employed in order to run slightly downhill over such a long distance. I looked in vain for any sign of the tremendous excavation that must of taken place. Or did they bore a tunnel? If so, that must have been a big powerful machine. The engineering feats of the 20th Century are a wonder to behold for their sheer scale and complexity. The long-term costs of these projects, though, have not been so wonderful.

So much of the land called "wild" in the western United States of America is like this: rammed through, cut clear or ridden over. The particular parcel where I started my birthday is managed by the Bureau of Land Management and is promoted as a recreation spot for OHVs (off-highway vehicles). Signage instructed riders to stay on marked trails due to the area being habitat for the endangered Desert Tortoise. Poor *Gopherus agassizii*, who survived a dramatic—though gradual—change in climate that transformed a tropical forest into a desert, only to have its burrows crushed by off-roaders. Of course, the not-so-gradual Climate Change of our own era will likely drive this creature to extinction. Such are the values of our society that unmonitored destructive activity is a legally-sanctioned and culturally acceptable use of "public" land.

If you don't mind digging a hole for your toilet in the morning, BLM land can be a great place to camp for free. As with the National Forests, you are allowed to stay overnight just about anywhere (for a limited number of days) as long as you are not blocking any roads. I had accepted the birthday-honor of choosing the night's camping spot and had picked this one because it was desert, which we would soon be leaving. Clarabelle and I were on our way to Portland, Oregon, from Joshua Tree, California, by way of US Route 395, which starts near Hesperia, California, ends at the Canadian border, and passes through many truly scenic areas on the way.

We had camped on the edge of a place called Freeman Canyon. The terrain was rocky but softened by greasy-leaved, winnowy-branched Creosote Bush (which can form clonal colonies several thousand years old), grey-green, tufty-shrubbed Rabbit Bush—brightly adorned with summer's first blossoms of golden yellow—and frilly carpets of desiccated *Cryptantha*, its ephemeral flowers long spent from their brief exclamation. Large black ants filed back and forth from their home in a column a foot-wide, collecting seeds.

The Sierra Nevada range was announced by a dramatic multi-lobed outcropping of stone protruding from the landscape like bone through flesh. It was quiet except for the wind. No OHVs or anything else disturbed the home turf of our threatened, shelled friend at that moment.

After breakfast, we hit the road but did not leave the environs of the aqueduct, which shadows Route 395 for nearly 200 miles from that point north. Soon we were driving alongside the site of the former Owens Lake, which was once a large body of water (12 miles long by 8 miles wide), fed by the Owens River. In 1913, the Los Angeles Department of Water and Power (LADWP) re-routed the river into the brand new Los Angeles Aqueduct. A little more than a decade later, the lake was almost entirely desiccated. Since then, groundwater pumping by the LADWP to supplement the take from the river has lowered the surrounding water table enough to dry up other seeps and springs that nourished the area.

The destruction of Owens Lake and the lower Owens River was—and is—an ecological disaster. Previously, millions of migratory birds had used it as a vital stop-over, but their habitat was drastically reduced. Meadows of native flora—which early European visitors described as filling the valley floor as far as the eye could see—have desertified. The lake bed's exposed alkali soil, whipped up by the wind, is the single largest source of dust pollution in the United States according to the EPA, and contains toxic constituents such as cadmium, chromium, chlorine and iron that are harmful to breathe.

The construction of the aqueduct sparked what are famously known as the "California Water Wars," which set Owens Valley farmers against the LADWP. The LADWP's bad behavior began even before the aqueduct was built, when they sent agents to the area posing as farmers who bought up as many parcels as they could in order to gain the water rights for the utility. In the 1920's and 1930's, Owens Valley farmers actually dynamited portions of the canal to provide themselves with irrigation water. The conflict is often styled, sometimes romantically, as the classic country vs. city clash —of simple agricultural folk struggling to survive an assault by effete urban hedonists—but this simplistic treatment ignores certain historical realities of the area: namely, its invasion and conquest by Europeans in the mid-1800's.

Previously, the Owens River valley was the home to semi-nomadic Native Americans, specifically the Paiutes. They called the river, Wakopee, and the lake, Pacheta. They hunted, gathered, and practiced agriculture. Their diet included Wild Hyacinth tubers, yellow Nutgrass corms, grass seeds, and pinenuts. They hunted deer, desert big horn sheep, small game and caught fish. They also collected the pupae of the Alkali Fly (*Ephydra hyans*) an insect that laid its eggs on the surface of the lake; the pupae would build up along the shores where they were easily collected and then dried for storage.

The fact that the Paiutes built a ditch irrigation system for watering certain of their crops has been cited to justify the use of canals by ranchers; however, the methods differed in purpose, operation and scale. The Paiute used their system to expand the range of native plants they ate, not to support imported animals or crops. Dams built on the Wakopee were often seasonal constructions, in place for only a portion of the year, and any fish stranded in the channel by the lowered water flow were conscientiously harvested: a philosophy of wasting nothing. Conversely, modern ranchers make permanent diversions, ignore the effects they are having downstream and are wasteful. A common sight on ranched lands, including the the Owens Valley, is the use of large sprinkler systems spitting out streams of water in the middle of the day; water loss through evaporation under those circumstances can be as high as 50%.

Among the first European visitors to the Owens Valley were "mountain men" Jedediah Smith, in 1836, and Joseph Walker, in 1834. In 1845, John C.

Fremont, a military officer and later the first Republican candidate for President, led the first full-fledged expedition into the area. Fremont named the valley, river and lake after his compatriot, Richard Owens, who had served with him during the seizure of *Alta California* from Mexico. Not once in his life did Owens ever see the area. (Here is yet another opportunity for re-naming as decolonization.)

Until 1859, few Europeans visited, but in that year a military force led by a Captain John W. Davidson was organized and sent there for the nominal purpose of recovering stolen horses from Native Americans. When they met the Paiutes, Davidson was impressed, describing them as "interesting, peaceful, [and] industrious." Furthermore, the Paiutes had not approved of the horse-theft and had already disciplined the perpetrators, some with the penalty of death.

On this expedition no blood was shed and everyone parted on good terms. Davidson pledged to the Paiutes that "so long as they were peaceful and honest the government would protect them in the enjoyment of their rights." The Paiutes replied that "such had always been their conduct and should ever be—that they had depended on their own unaided resources —that they had at all times treated the whites in a friendly manner and intended to do so in the future." They also promised that anyone in their tribe who broke this word would be "punished with the sword." Davidson closed by saying to the interpreter: "Tell him that we fear it not, that what I have said I have said. I have lain my heart at his feet; let him look at it."[1]

Everything went downhill from there. Davidson's attempt to set aside the area as a reservation failed when Congress did not pass the needed legislation. (Not that this would have guaranteed protection for the Paiutes, as the history of broken US treaty obligations with Native Americans illustrates all too well.) Meanwhile, cattlemen and prospectors started arriving in the area, their numbers swelling quickly. The winter of 1861-62 was severe, bringing suffering to the Paiutes due to the lack of game to hunt the following spring. Conflicts with the settlers began when a Paiute killed a steer that was grazing on their Hyacinth fields.

The European settlers had little or no respect for the rights and lifestyles of the indigenous people and soon the situation escalated to bloodshed. In mid-1862, the settlers asked for, and got, a US military force sent to the area to bolster their side. By 1863, the Paiutes had been subdued. Owens Valley now belonged to the invaders, who were predominantly ranchers.

Only fifteen years later, in 1878, excessive irrigation by the ranchers caused the water levels of Owens Lake to start dropping. In other words, their methods were already unsustainable before the arrival of the aqueduct, a fact now little remembered. The plain truth is that, lake or no, the arid Owens Valley is not an appropriate place to raise cattle. That ranching is allowed to remain there at all is in defiance of all logic (except that of the short-term financial variety).

Ecological issues entered the "Water Wars" in the 1970's when Inyo County sued Los Angeles under the terms of the California Environmental Quality Act. In 1997, two nominally environmental groups, the Sierra Club and the Owens Valley Committee, were parties to the Memorandum of Understanding that required the LADWP to allow some water to flow into the river and thence to the lake. Over the decade that followed, a series of court orders forced the LADWP to make good on its promises.

In the present day, Climate Change is now an exacerbating factor, threatening to throw the fruits of these hard-won battles into the garbage. Witness: "Owens Valley ranchers and environmentalists brought together by drought," a front-page article in the L.A. *Times* on May 20, 2015, published shortly before our journey to the area.

A quick summary of the story: Due to the prolonged drought in California, and specifically the record low snow-pack in the Sierras, the L.A. Department of Water and Power has declared that it can not deliver promised (and legally agreed upon) water allotments for both the restoration of Owens Lake and the ranching industry. Parties calling themselves "environmentalists" have caved under some implied but unspecified pressure to give up a portion of the restoration allotment so the ranchers will not have to lose as much.

That sounds simple enough, but to make full sense of it a reader needs more: clear differentiation of opinions from facts, some "follow-the-money" analysis, and a decent helping of "big picture" contextualization. This is not what the author of the piece, Louis Sahagun, delivered. As is typical of the corporate media, the article fell far short of telling the whole story, took the side of the moneyed status quo, and played loose with the truth—all while speaking in a tone falsely conveying "objectivity." The modern newspaper does little more than construct a "he said/she said" narrative around every issue without questioning the veracity of what is being said and calls it "balance." Investigatory work is not only expensive but likely to upset advertisers. This was the predictable (and predicted) result of Clinton-era legislation deregulating media ownership. Too few people keep any of this in mind when reading the corporate media.

A lack of clarity begins with the very first teaser line, which is placed at the top as a quotation to tweet: "If ranches go dry, owners will lose livestock as the *natural habitat* on the property succumbs to drought [my emphasis]." Since when is irrigated pasture "natural habitat"? Later in the article, this very point is called out: "'A lot of people think all this green grass is natural,' Talbot said, while inspecting the latticework of creek-fed ditches in his browning pastures. 'Without irrigation, it'd be nothing but sage flats.'" So are "environmentalists" calling irrigated pasture "natural habitat"? Apparently so: "Environmentalists say the loss of habitat [on ranches] would be disastrous to wildlife and vegetation in the valley." But sage flats do not need irrigation so the mystery is not solved because the article says no more. Fortunately, a helpful commenter who sounded like a knowledgeable local noted that certain irrigated areas "are in the historic wetland and meadow locations that have existed here for millennia." That is indeed helpful to know, but does it makes ecological sense to keep the cows there? No.

This point is made splendidly by George Wuerthner and Mollie Matteson in their highly informative book, "Welfare Ranching: The Subsidized Destruction of the American West":

> *Perhaps the biggest fallacy perpetrated by the livestock industry is the idea that if we would only reform or modify management practices, there would be room both for livestock and for fully functional ecosystems, native wildlife, clean water, and so on. Unfortunately, even to approach meaningful reform, more intensive management is needed, and such management adds considerably to the costs of operation. More fencing, more water development, more employees to ride the range: whatever the suggested solution, it always requires more money. Given the low productivity of the western landscape, the marginal nature of most western livestock operations, and the growing global competition in meat production, any increase in operational costs cannot be justified or absorbed*

> *Even if mitigation were economically feasible, we would still be allotting a large percentage of our landscape and resources—including space, water, and forage—to livestock. If grass is going into the belly of a cow, there's that much less grass available to feed wild creatures, from grasshoppers to elk. If water is being drained from a river to grow hay, there's that much less water to support fish, snails, and a host of other life forms.* The mere presence of livestock diminishes the native biodiversity [my emphasis].[2]

Sahagun betrays an anti-scientific bias with his single statement about the ecologically deleterious effects of ranching: "Environmentalists have long *believed* the local mountains would be better off without cattle trampling stream banks, polluting creeks with animal waste and eroding fragile meadows with intensive grazing" [my emphasis]. Oh, this is just something that "environmentalists" "believe," is it? It's not a "belief" that fragile desert ecosystems and riparian zones would be better off without cows; this has been exhaustively documented over the course of many decades in a multitude of locations. You can read entire books on the subject. But if Sahagun acknowledged these facts, it would muddy the scripted ranchers-vs.-environmentalists squabble.

The bottom line, of course, is... the bottom line. The article states that, according to "officials," "Farming and ranching generate $20 million a year in Inyo County." This—and the sepia-toned myth of the self-reliant cowboy (which is a total fallacy)—is why no one can raise the fundamental question of whether ranching should even be allowed.

And it shouldn't be. In the big picture, $20 million is chump change. If we were a rational society, we would just pay that—or double the amount—to make it stop. It's a small price for ceasing such destructive activity. Of course, we should also shut down the aqueduct and deed all LADWP land back to the Paiutes. They were far better stewards.

The first town you hit going north on the 395 after the former Owens Lake is Lone Pine, home to a movie museum celebrating the area's frequent use as a backdrop for Hollywood productions. The scenery is quite striking, with the 14,505 ft Mount Whitney dominating the western view. After traveling through desert for hundreds of miles, the town is unexpectedly verdant, with many trees and irrigated pastures. The city park where we stopped to have lunch was centered around a winding stream with giant cottonwoods and a thick lawn of grass and clover. People were enjoying the cool shade on that hot day: families with children, tourists with picnics, and the local home-bum set, who were blazing up joints and laughing loudly in a back corner. Dogs invariably gravitated to the stream for a splash. The sound of the wind in the foliage of the Cottonwoods was like a whispering crowd. The Romans noticed this quality many centuries ago and named the tree *populus*, their word for "people."

After lunch, Clarabelle and I were walking through the park, admiring the trees, when we found two baby birds at the foot of one particularly large *Populus*. They didn't look old enough to be on their own. Indeed, everything about their demeanor—their downy feathers, their awkward movements, and most of all their plaintive cries, emitting from mouths that seemed over-sized for their stature—suggested that they were out of the nest prematurely. One had a bit of runniness in its eyes and was carrying a wing not quite right.

We were both immediately filled with compassion and doubt. We felt so bad for them but... but what could we do? We crouched down with them for a few minutes to take better note of their conditions and behavior. Were they in danger? Was there any way we could help?

We called an animal-loving friend who knows about these things and he told us they didn't stand a chance of surviving.

We hung around a little longer, watching them, then finally decided to leave, sadly. But we sat in the van for a long moment after we started the engine because just then a couple with an unleashed dog was approaching the spot. We kept our fingers crossed that our little avian friends weren't about to become canine food. They didn't. The people and dog moved on. So did we, but not without reservations.

For the next few miles we agonized over whether we should have taken them with us. We could have put them in a box and fed them worms from the bait shop, we surmised. But did they need their food pre-macerated at this age? Neither of us wanted to chew on night crawlers. The highway was divided by a barrier in that section but an opening was coming up where we could turn around and go back. "Should we?" one of us asked, but the other didn't answer and we kept driving.

Quite a few people would say we were stupid to worry. After all, this is just nature. Some babies live and some don't. Maybe the mama bird pushed them out of the nest on purpose. "Survival of the fittest." We could claim that our emotional response sprang from sensitivity, but a "realist" would counter that we were only indulging a sentimentality born of disconnected city life, nothing but wallowing in kitsch.

I am utterly familiar with all the reasons it "makes no sense" for me to care the way I do about non-human living creatures, especially injured ones. The let-'em-die rationales are trumpeted with a multitude of fanfares, from the dick-swinging impudence of our hyper-macho culture, to the presumptuous claims of "dominion over creation" by our patriarchal religions (the Pope's eco-posturing notwithstanding), to the rapacious devouring of nature's bounty by our corporate overlords (while they greenwash all the way to the bank). To put it simply, our society is just *mean*. Was our heartache for the baby birds an overcompensation for all of this, understandable but pointless? I don't know. But what I can say was that the overwhelming sensation was one of helplessness: of not knowing what to do.

* * *

Along the 395 are signs for a turn-off to see a Bristlecone Pine forest. Among the oldest living things on earth, *Pinus longaeva* individuals can reach ages in excess of 5,000 years. We took the turn, onto California State Route 168, and soon we were climbing up the Westgard Pass alongside a dry stream bed.

After a couple-three miles we stopped to check out some Prickly Poppies blooming in the channel. Both being plant lovers with a particular interest in wild medicinals, we had to stop to observe what turned out to be *Argemone munita* in its natural habitat. Their name is apropos, as the leaves are sharply-toothed and the pollinated fruits covered with spines. The showy blossoms, swaying in the breeze, were comprised of papery-white petals arranged around a bulbous-headed style (the female part) in the midst of a riotous crowd of orange anthers (the male parts). Yellow pollen dusted the petals and every insect that visited them.

Prickly Poppies have a long history of medicinal and ceremonial use by Native Americans. Nowadays, they are popular on "legal high" websites due to their euphoric opiate effects. We went from plant to plant, checking them out at every stage of blooming, from fresh-as-a-sheet to raggedy.

When we turned our attention to our greater surroundings again, we noticed that big, dark clouds had broken free from the Sierras and were piling up on top of the peaks of the White Mountains above us. We still had twenty miles to go, up onto ridges that were now coming under deepening shadow. Having been caught in a mountain thunderstorm before—and having the living daylights scared out of us by close lightning strikes—we decided to put off our trip to the Bristlecone Pines again, and head back to the valley and continue north.

As we were approaching the bottom of the slope, but still had a wide view of the landscape below us, we saw a cloud of dust off the side of the highway about a half mile ahead. We wondered if it was a dust devil or maybe someone off-roading, but it stopped before we could make a positive ID. About 90 seconds later, the mystery was solved. A vehicle was parked on the shoulder and the driver was waving his arms for our attention. We pulled up and he asked urgently if we had a phone and pointed to the ditch off on the other side of the road. There we saw what had caused the disturbance: a motorcycle was laying on its side and, a few feet away, the body of a woman in riding gear, face-down and unmoving.

My traveling companion called 911 immediately. Unlike with the baby birds, there was something we could do, and that was it. The dispatcher got our information and asked us to wait until help arrived. We approached the body.

It was a woman in riding leathers, the patch on the back of her jacket proclaiming her affiliation with a Harley Davidson club. Her hands were pinned underneath her body and her ponytail was flipped up over her helmet. Her face was in a pool of blood in the dirt. Clarabelle went to one side and I to the other.

"She's still breathing," said my friend, and indeed I could hear her labored gurgles. My friend stroked her arm, saying, "Keep breathing, sweetheart. Just keep breathing." I put my hand gently on her back, doing what I thought I might want someone to do for me if I was the one laying there. I could barely feel the rise and fall of her body through the thick leather. Looking at her more closely, I estimated her age to be right around my own, which was certainly a sobering observation on my birthday.

The accident had been serious. The motorcycle was facing the opposite direction that she had been traveling. All her gear was scattered around a fairly wide area. I saw a sleeping roll, an iPhone, a tube of Carmex.

We hoped she was not in pain, and probably she was not, "shock" being what it is. I hoped that she was not afraid but I could not tell if she was conscious. Clarabelle kept softly saying, "sweetheart." It seemed like an eternity waiting for the paramedics.

A flash of paranoia hit me. Could I "get in trouble" if the paramedics saw me touching her when they arrived? Would the family sue me if they believed I had somehow hurt her worse? With these questions, the paranoia was joined by resentment: what a frightful and frightfully legalistic society we live in that I was even having such thoughts. Must we fear to comfort the apparently dying? Nevertheless, when at last we heard sirens on the highway, I kept an eye out and stepped away well before the emergency vehicle and its team arrived.

As it turns out I was the last of their concerns, and they ignored me as they sprang into action. It was a fire crew made up of one middle-aged man and two younger women. The two women seemed a little shocked at the sight, but the man betrayed no emotion other than surprise that the injured biker was still breathing. They turned her over, put a tube down her throat, and focused on their jobs.

I spoke briefly with the driver who had flagged us down. He passed a few quick words with his companions in a language I didn't know so I asked him what it was. "Spanish," he replied. I expressed surprise that I hadn't recognized it. "We are Argentinian," he said and we both smiled. He knew their accent was quite foreign from the "Mexican" Spanish frequently heard in the US. He seemed ready to go so I mentioned that the 911 operator had asked witnesses to stay on the scene until the police arrived. He agreed to do so, but was clearly a little nervous about it. His companions looked terrified. The news had recently featured coverage of the murder of Freddie Gray by Baltimore police officers and the protests that followed, so I could totally understand why they might be hesitant to interact with American cops.

Before the woman had been loaded into the ambulance the California Highway Patrol arrived. One officer checked in with the fire crew, two started inspecting the scene and the fourth approached me and the Argentinian. Having established that none of us had been present at the spot the moment the accident occurred, he told us that he was asking us to leave. Then, he looked me right in the eye, but without the cold menace that cops usually project, and put out his hand to shake mine. "Thank you," he said, with utmost sincerity. "Most people don't stop."

"They don't?" asked Clarabelle, who had just joined us.

"No," said the cop.

"I can't believe that," I said.

"It's true," the cop replied. "So, thank you."

We turned and left. As soon as I was seated in the van, I burst into tears and cried hard. I was sad for the injured woman, but also to be living in a place where "most people don't stop."

We drove the rest of the way down the 168 to the 395, but instead of getting back on, we turned off into a parking area with an historical marker to sit and breathe for a moment. The whole time we were at the accident scene, different vehicles drove by, some pausing to see what was going on. One car in particular had caught my eye: a Model T driven by a white-haired white man. He pulled up beside us in the parking area now and asked us about what we had seen, but didn't listen much, giving us his theories about what happened instead. "He must've been going 85 to go off the road like that," he claimed, implying reckless behavior on the part of the biker.

"She," Clarabelle said, but the man ignored her, and kept going. "She," she repeated, in response to another usage of "he," but the speaker wasn't paying attention, as he tried to build a case for negligence or even recklessness. Eventually Clarabelle cut him short, though courteously, and he drove off.

"Do you believe that?" she asked.

"Playing the blame-the-victim game," I said.

"Like he knows what happened." She was clearly flustered.

The Model T was so emblematic of his father-knows-best haughtiness that I could hardly believe it was real. The anachronistic vehicle and the old-fashioned moralizing were such a perfect match that I knew I would have to write up this story like I have here, as a true-life narrative; in a work of fiction, readers either wouldn't "believe" a detail like that or would (rightly) find the symbolism trite. Yet there it was.

I stepped out of the van to read the historical marker. It was set underneath a Giant Sequoia tree (*Sequoiadendron giganteum*), which is not native to that area though they grow nearby in the Sierras.

The marker identified it as "The Roosevelt Tree," planted July 23, 1913. At 101 years old, it potentially has centuries of life before it. The inscription said:

> *This Giant Sequoia tree is*
> *reported to have been*
> *planted to commemorate*
> *the opening of Westgaard* [sic]
> *Pass to automobile traffic.*
> *The tree was named*
> *in honor of President*
> *Theodore "Teddy" Roosevelt.*
> *Presented by the Big Pine Civic Club*

How proud people were of their wilderness-breaking infrastructure projects in the 20th Century! On the day the tree was planted, did they wish for the safety of all who would use the road? Maybe. Did they foresee that the next century would bring 3,599,489 motor vehicle deaths in the US? Doubtful. How would they have reacted if a time/space wormhole had opened and showed them the woman biker on the side of the road, trying to breathe in a puddle of her own blood? Would it have broken the spell? Maybe for the people who witnessed it, but the road was already built, and what should they have done then? Dynamite it?

In our own time, scientists are giving us very clear pictures of the quite dire future we are bringing on with our current choices. How are we reacting? California Governor Jerry "not-a-moonbeam-now" Brown is requiring California citizens to cut their water usage, but is letting corporate agriculture and the fracking industry have their wasteful fill. US President Barack O-bomb-ya is allowing Royal Dutch Shell to drill for oil in the Arctic, even though Shell's own study shows that such activity will contribute to global warming.[3] And at the former Lake Owens, the short-term bottom-line of a small class of welfare-dependent businessmen (aka "ranchers") is benefiting at the expense of the environment. We are certainly no wiser than the road-builders of 1913, and I would wager that we have become less so.

Speaking of dynamite, Derrick Jensen has written that if you want to make a difference, blow up a dam.[4] Positive environmental effects would result, especially for salmon, and—unlike with the Los Angeles Aqueduct in the 1920's—there's a high likelihood the infrastructure wouldn't be fixed. At what point does it become our existential imperative to devote ourselves to such meaningful actions instead of talk, half-measures and politicking? For those species that have already been extinguished from the planet during what can no longer be denied is the Sixth Great Extinction,[5] that point has passed.

* * *

Back on the 395, we headed toward Mammoth Lake, there to meet a friend from Portland who was hiking the Pacific Crest Trail and happened to be staying there that night in a hostel. On the way, we stopped at a pull-out to check the oil (which you can't do too often with '80s-era Toyotas). I found *Cryptantha* flowers in bloom, unlike at our camping spot that morning where they were already finished for the season. By traveling farther north and higher in elevation, we had effectively gone back in time.

Cryptantha is Greek for "hidden flower," which aptly characterizes their tiny size. Each blossom was perched on the end of a long, fuzzy green ovary, and the plant's spindly stems were bent in the breeze like the branches of an old pine on a mountain ridge.

Cryptantha is an example of an "ephemeral," a quick-growing annual that sprouts, flowers and seeds in just a few weeks. Individuals of our species, *Homo sapiens*, regularly survive for decades—longer or shorter on average depending on where we live—but that is "ephemeral" to a Creosote Bush clonal colony, a Bristlecone Pine or a Giant Sequoia.

* * *

We wondered about the woman biker every day after that, for the rest of the trip. Back in Portland, we looked up the accident and found out that she died about two hours after the crash. She was 48, two years older than me. If her Facebook page is any measurement, she was loved and appreciated by the people in her life, who included a daughter and many Harley riders. She was commonly described by posters as a "sweetheart," so I guess Clarabelle called it right.

We also learned that someone would have helped the birds: Eastern Sierra Wildlife Care.

July 1, 2015

Notes:

[1]   Guinn, J.M. "Some Early History of the Owens River Valley." (Annual
      Publication of the Historical Society of Southern California, Vol. 10, No. 3 (1917)),
      pp. 41-47.
[2]   Wuerthner, George and Matteson, Mollie. Welfare Ranching: The Subsidized
      Destruction of the American West (Foundations for Deep Ecology 2; First
      edition, August 1, 2002).
[3]   Macalister, Terry. "Shell accused of strategy risking catastrophic Climate
      Change" (*The Guardian*, 17 May 2015).
[4]   Jensen, Derrick. "Actions Speak Louder Than Words" (1998).
      http://theanarchistlibrary.org/library/derrick-jensen-actions-speak-louder-
      than-words [retrieved 7/8/19]
[5]   Ceballos, Gerardo, et al. "Accelerated modern human–induced species losses:
      Entering the sixth mass extinction" (*Science Advances*, 19 Jun 2015: Vol. 1, no. 5).

"No Enemy to Conquer"— Discussing Wildtending with Nikki Hill

Nikki Hill has been pursuing an interest and practice in wildtending since 2012. She holds a degree in Environmental Science with a minor in Botany, but her knowledge on this subject has all been gleaned from field work and mentors, not academia. We crossed paths in Mendocino County as I was finishing this book and sat down to talk about some of what she has learned and observed on her travels.

KOLLIBRI TERRE SONNENBLUME: How do you define "wildtending"?

NIKKI HILL: Wildtending is a relationship with the land, the plants, the animals, in which you are tending for increased abundance and health.

KOLLIBRI: "Tending" being particular activities?

NIKKI: Yeah, particular activities. The way in which you're interacting with the world. So if you're out harvesting things, you're harvesting at a proper time that would be beneficial for increasing the abundance of that plant later, rather than decreasing. Or you're doing it at a time when things are seeding. Or you are burying canes to make more shrubs if it's berry bushes. Those are examples.

KOLLIBRI: So you're doing things to the landscape directly?

NIKKI: Yes, it's a very human-level, individual-level activity.

KOLLIBRI: And community level too?

NIKKI: Community level too, with all the individuals involved having that focus, that understanding and that relating.

KOLLIBRI: Examples of people who have engaged in wildtending are Native Americans on the North American continent, in the western half, specifically.

NIKKI: Yes. A lot of nomadic, hunter-gatherer tribes.

KOLLIBRI: How is wildtending different than what we think of as hunting-gathering?

NIKKI: The tending part comes in where people were not just hunting and gathering things. The tending part is all the other activities they were doing so they had something to hunt and to gather. Like planting seeds. It's not often talked about that I know of, carrying seeds—even in reading different stories about how cultures and different tribes operated—but it was a big part of people's lives.

KOLLIBRI: What you're saying is that people weren't just harvesting, for example, berries, and eating them; they were planting berries too?

NIKKI: Yes. In different ways. There was just this attention and understanding that your actions have an effect. So maybe you're eating a bunch of berries and then you're pooping in a place where you know they'll grow. Simple things like that. Other times you are actually making an effort to collect seeds and carry them to another place where there's not the foods you would like. So you're planting new gardens and spreading the plants that way. So that wherever you're walking in your life, there is food for you to eat. You're increasing that abundance.

An example of what we see of that today is the Amazon rain forest. There's different researchers who have come to see that parts of the Amazon were actually planted gardens. They're a food forest, a very old one that we consider wilderness today. That's part of what I've been learning here in the Great Basin, in the West, is that what we consider wilderness there was actually from centuries of planting. From people walking around with these seeds, with this awareness, with this relationship of increasing abundance of the things they relied on, in the places that they walked. Today we see that as wilderness. The fabric of our Great Basin desert had an influence from people. A larger influence than people [now] understand.

KOLLIBRI: Right. Charles Mann, the guy who wrote *1491* and *1493*, talks about this. In the Midwest, the prairies were expanded to enlarge the Buffalo's range by having regular fires. So they were working not just on the individual plants but entire landscapes. That's wildtending too.

NIKKI: Absolutely. That's one you hear more of. People talk about fire a lot. It's a larger scale kind of picture. In some ways that one gets a lot of focus, and I'm not saying it shouldn't, but it's familiar to be doing things on a large scale. But I still think it was smaller scale than a lot of "management" projects we see today. Where we come up with an idea about what's good for one part of forest and that becomes policy. And then that's done throughout all the forests whether or not it makes sense, even in that region. There's some sort of mentality about that scale that we're still working with. But yeah, fire was definitely an example of wildtending.

KOLLIBRI: And then moving plants from one area to another?

NIKKI: Yes. That's more where the day-to-day relationship comes in because it's not like it was on a massive scale all at once, like you're tilling a field and then planting. It was happening just as a way of walking, as a part of your movement, you were carrying seeds and planting them. You're digging roots and you're planting seeds.

KOLLIBRI: Yampah, is such a good example of what wildtending is about. What happens with that plant over the course of the year?

NIKKI: There's two different kinds: an early Yampah and a late Yampah. The early Yampah comes up in grassy areas or rolling sagebrush hills, often in a place that at some point of time in the year there is some standing water. [It's] not a bog habitat that she's growing in but a place where there's been some standing water, like a flatter place in the rolling hills, [sometimes] with mounds around the edges—I was seeing that this summer. It comes up in the spring and you can start digging roots at any time of year. Hopefully, if you're digging, you have some seeds to put back in place. You find a large patch and start digging from the center, is how I've been doing it. A lot of this is finding my own way. I'll go to the center of the patch and dig some roots there, thinking that the edges of the patches, that's where things will be seeding and expanding that garden [on their own]. So I want to keep as many plants as will seed that year on the edges. The flowers flower and then later in the summer, the seeds are ripe. That's the time of year that's really beautiful to be harvesting roots because as you're harvesting, the seeds are ready to fall. You're picking a place where the roots are dense so that you can make one hole and get a bunch of roots out of it. Then when you're filling your holes, you're not necessarily making it look like you haven't been there. You want to leave cracks open for those seeds to fall into, en masse. Part of what you're doing is you're planting like you would want to dig, in densities. Because it's more work to be digging these wild plants in the rocky soil than to be in a well-tilled agricultural field. So it's nice to have a place where you're getting lots of roots from one hole. So your harvesting and planting actions are done in such a way to encourage that.

KOLLIBRI: You showed me a place once where the rocks appeared to have been placed on the ground in such a way as to easily lift them away and harvest.

NIKKI: People were figuring out, or probably were very sophisticated in figuring out, that some of these roots—not the Yampah, but some of these *Lomatium* and *Cymopterus* roots in the Carrot family—they grow really long tap roots that are hard to dig. Some of them go down a foot or more and they're kind of skinny until you get to the bottom, so you really need to try dig them out. So they figured out if they had these trenches from their digging already, if they put down the rocks like puzzle pieces, when you go back, you just pull back a rock, the root will be right there. So the stories I heard about Shoshone aunties who planted this way were about Big White Coush, this big-rooted *Lomatium*. Instead of growing straight down she would hit the rocks and start to grow sideways and start to swell so they would be digging up "loaves of bread"—it was called "Bread Root." So it shows a very intimate understanding, responding to how those plants grow and working with them to make it easier.

KOLLIBRI: This is different from farming, obviously, even though they were planting seeds.

NIKKI: Yes.

KOLLIBRI: Because they're working with wild plants, not domesticated plants.

NIKKI: With wild plants, you're stepping into this relating where you're not just trying to dominate how they grow; you're aiding a process that's already wanting to happen. You're working *with* it. They weren't out there clearing spaces in the desert and then planting what they wanted. They were planting where it made sense, in the places that were there. These were all long-lived perennial plants. There's a difference there. Most of our agriculture today is based on annual foods. That's a pretty big transition from subsisting on perennials. Some of these roots out there were very long-lived. You'd be harvesting them when they were fifteen or twenty years old for good-sized roots. Some of them live for up to a hundred years. So there's something with the nutrition there: plants growing in these mineral soils, not irrigated; they're going to be concentrated in different way than an annual would, which doesn't have the chance to do that. [With agriculture] there's more watering that's diluting minerals. Also with these wild plantings, you're planting for your grandchildren. That's a pretty different relationship and spirit, and interaction as well. There's no instant gratification. You're stepping into being part of a process more than you are with farming because you're not spending all your time on these gardens making sure that they're growing right either. Something happens—a drought or different insects come through that season—you're not there for it. You're participating in chance. With a spirit of encouragement, encouraging these things. Most of the time they're out there having their own life. A farm field does not have its own life. The plants there are having their weed friends who come in taken away from them, and artificial nutrition—whether organic or synthetic—being sprayed on them. It's a different energy. [With wild plants] you're not pushing them the same way. You may be finding ways in which you can tweak how you're setting it up—like the bread root and the rocks—but you're not insisting that it be a certain way.

KOLLIBRI: This started as a Native American way of getting along. Obviously it probably was a way of getting along in other places in the world too for a long time, but here it was done until pretty recently. Like, there were tribes out here who didn't get conquered until the late 1800s.

NIKKI: Yeah and there's still tribes participating in these activities. I understand why they wouldn't be talking about it much, but it's still happening, it's still a part of their cultures. I've been to some pow-wows and different ceremonies I've been invited to go to, and a lot of the ceremonies are around their food. It was central to their lives, this relationship with their food.

There's places in southern Oregon that have some of the densest gardens I've seen across the West and I've traveled from Oregon down into California and I've spent some time in Nevada, New Mexico and Colorado. I've been appreciating the education I've been getting—the picture of how these plants walk, their range, and to imagine how the gardens must have been like when they were more intact. You find these fragments and it's just mind-blowing. These places in southeastern Oregon—these dense gardens—there was still tending going on there until recently so they're less fragmented. The density shows that the interaction there was not long ago. It was definitely done with intention.

KOLLIBRI: You've been using the word "gardens" to describe these places.

NIKKI: Yes. That's what I started to see them as. That's how they were pointed out to me. Standing in the middle of the Great Basin you're looking and maybe you start to see some slight mounds in the landscape and someone tells you that's where the winter camp used to be. That's where the tipis were for many months. Then you go around those areas and the plants are very dense. As you walk around the area, you start to get this picture. It's different than a field.

They aren't seen as gardens. It's seen as wilderness or just how the desert is, but they really are. They really were gardens. [But for someone] not recognizing that they were gardens, they don't notice when they start to disappear in places because there's no longer that relationship happening. So when people talk about Great Basin habitat or range-land, they're not speaking of these plants. They don't know them. They don't understand that they were in gardens, or that these gardens were part of the fabric that made the landscape—which people are talking about conserving—but they don't understand the individual players who are there or that there was a relationship going on there...

There's still an agricultural mentality with our approach to restoration actually. We have this idea that we need to clear something to make room for the things we think should be there. I think that's a fallacy. That's part of why we don't see more planting projects. With restoration, one of the first thing you apply for with grants is eradication. It's not working. You can talk to different people who have been in the field and they're exasperated. I talked to this Forest Service guy in Nevada last year and he knows it's not working. But that's the policy is the first thing you do is eradication. That's the bureaucracy part.

The other part is the mentality that we need to clear the way first but I think that goes against what nature is showing us all the time. That's not how it's doing it.

KOLLIBRI: You're talking about the difference between doing something *to* nature versus doing something *with* nature.

NIKKI: Yes. Dominating versus participating.

KOLLIBRI: It seems like our culture wouldn't really know what to do with that phrase, "*with* nature."

NIKKI: Yeah and that's the part where you just have to be there. That's the part. Participating means you need to suspend your ideas that you've been given. A lot of people have ideas about how things should be done, and they're not their own ideas. They're part of the societal vernacular that you pick up.

This is different than that. You just have to actually be there and participate with it. If we start doing that— the different individuals on these different projects, or just on our own with our own curiosity—then I think there's a chance for some new understandings to happen about what it is to actually participate. But if we don't start to try to do that, we'll never know. If we continue with an outside way of approaching things, it will continue to be top-down because of what we're familiar with.

So that's a part of what I like about the wildtending: it's an individual pursuit of curiosity as well. You start to understand more, the more you do it. Like, I'm interested in this thing becoming more abundant, or seeing more diversity here. What's the way I can work with what's already here? Trying to understand what it means to participate rather than dominate, where we are constantly creating an enemy. That's a part of that dominating mindset, is that there's always an enemy to conquer. There's no enemy to conquer, just a process to work with.

KOLLIBRI: In agriculture in general there's enemies. If you're growing vegetables, your enemies are the weeds. If you're a rancher, your enemies are the wolves, or so you think. There's an adversarial base to agricultural systems.

NIKKI: That part really needs to be dropped to participate.

KOLLIBRI: So as an individual working on trying to learn about wild-tending, you've had these things challenged in yourself.

NIKKI: Yes. I don't always know what to do. Sometimes I spend a lot of time doing nothing. And then I spend other parts of the time under-standing what it means to give myself permission to interact. Because the other thing I can see happening is that people who are well-intentioned about wanting to preserve or conserve nature or to care about it—they're terrified to interact with it. We have this belief that all we can do is be harmful to the world around us so we should have these wilderness places that people are kept out of. You hear this talk that maybe we should have some places that nobody is allowed to go except for scientists. I don't think that's going to help us. I think that what's being called for at this point in time—with different things going on in our world—is that re-con-nection of an actual relationship. In your relationship with the world around you, you need to spend some time just being with it, allowing yourself permission to interact with it in different ways and see how it re-sponds in small ways and that starts to show you things beyond an intel-lectual idea. And that process starts to unfold in part on its own. That's a part of participating. You're not fully directing that process. There's some methods or some general ideas of ways in which you can start to ap-proach interacting—like right time of digging—or understanding just how plants propagate. I've been focusing on that part so that's where I tend to talk from. "Oh, I can carry seeds around." I mean, there's such a joy in seeing seeds germinate. There's so much that is gifted to you from just participating in that. Even in your back alley, planting that peach pit and seeing it sprout the next year—there's something in that that's beyond an ideology, and that's the start of it. I don't think there's a plateau we get to where now we have the information, so now we know how to do things. It's not just about information. So part of the wildtending endeavor as an individual and as a collective is to get more into that space of being actu-ally present. Which is going to be continually changing. That doesn't mean you might not come up with some things that actually seem to work. But we won't get there by theorizing about it.

KOLLIBRI: What about Climate Change? Are you seeing it out there?

NIKKI: I do feel like I'm seeing things. I spent two different springs, last year and the year before, where I went to these gardens in Nevada and I did a tour across the whole state at a certain elevation, at certain gardens that hug the mountains there, in the foothills. I was with my mentor—my auntie—and she was showing me these places that are super dense gardens. The year we had been there had been a pretty wet winter so we were seeing lots of individual plants of this certain kind of Biscuitroot—this *Cymopterus*.

We went out at the same time the next year and we couldn't find hardly any plants across the whole state. We were traveling along the 50. It was kind of alarming. What we heard from people around there is that they had what is known as a "false spring" where temperatures warm up in February and plants break dormancy. This is our assumption, our theory, of what happened with these gardens is that they broke dormancy when the weather warmed in February, but then it got cold again so they went dormant again, and they're not going to come up again. They're not going to break dormancy twice in one season. So part of what we are interested in and concerned about is how long those plants will be able to handle those different cycles. At some point, the strong, long-lived root needs to photosynthesize and it needs a longer period of time in which to do that. So I feel like we've been seeing some shifts happening with different plants. I think individual plants are good indicators for that. I think they start to give us more pieces of the puzzle, with what's happening.

KOLLIBRI: What's part of the response to the climate changing, in what you're doing? Like, planting things in different places, or—?

NIKKI: Yes. I've been taking seeds from one region and moving them to another. Taking seeds from lower elevation and moving them to a higher elevation. Planting gardens on a north slope rather than a south slope, so plants maybe stay a little colder and won't be breaking dormancy during those false spring events. It's kind of like a giant puzzle, not knowing exactly what will work for certain plants but trying to get a better picture of that. So yeah, moving them around, which is I think something people have done forever. I think it's just a part of what humans do.

So, starting gardens with different varieties of Yampah, from three different states, and bringing them to the high mountain areas and seeing how they will do there. The other thought that goes along with that is that you're introducing different genetics into these different populations so that they have more genetic information to go from, to respond to changes. The idea is kind of a living seed bank. These plants growing in place, in different conditions, with more genetic opportunity, will have a better chance of adapting and being able to survive and continue with changing climate conditions than they would sitting in a seed vault. Scientists will call this "assisted migration" but it's a very old thing. I don't think it's very new at all.

KOLLIBRI: Although the changes that we're going through, the last twenty years, and what we're expecting in the next few, are more drastic than has occurred in the last 12,000 years.

NIKKI: Yes. So it seems like a good time to be walking with plants. I think the plants are showing us that they're doing those things anyways.

KOLLIBRI: Here in California that there's a ground squirrel or a chipmunk who is now found several hundred feet higher in the Sierras than it was before, because it's getting hotter and it wants to move up to where it's not so hot.

NIKKI: Life is already responding to these things. I think the plants are moving themselves around as well.

KOLLIBRI: But they can't move as fast; that's the problem. Not like a bird.

NIKKI: They can't move as fast. They have the birds to help them. Birds do a lot of planting, and other animals, but humans are part of that equation. They always have been. I think that people probably responded to these things in a similar way before, moving things around that they care about and that they wanted to continue, or that they understood the value of. So, the other part about the wildtending is that individual connection to individual plants, the things that you care about. That's a part of conservation. It's not abstract; it's very direct. You moving them around is part of how that happens. I think that shouldn't be left out of the equation when we're talking about these changes and what to do about them.

The part that's left out of the conversation all the time is that we don't question how agriculture moves all kinds of seeds around all the time, and plants into all kinds of areas that they're not from. Again, that points out that it's just something that people do. So how can we do it in a way that's more participatory? I think that's part of the lesson from the failing of agriculture that's happening, at least as we know it. It won't last being done in the same way. Too much domination. We can talk about moving our farms further north, but we should be moving the wild too...

They talk about how some seeds can lay dormant for fifty years. That's a beautiful adaptation and an incredible amount of intelligence—to know when the timing would be right—held in that seed. So that's another thing about participating is you're working with all these other intelligent beings that are very capable of responding to things as they are, and in that way we have a lot to learn from them. We need to come back into that place, which is not a place that's inaccessible to us. It's a part of who we are. It's just that we have this way of looking at things abstractly, with a mental process—and relying on that more than the response from our own being and our own senses—to tell us what's right and wrong in a given situation.

142

So I question myself out there, in these gardens. I want to make sure what makes sense. That it's okay. I'm not asking society that; I'm asking the places that I'm in. And it does respond to those questions, actually. I was planting cherry seeds along this river, and I was wondering if the spot was right for them or whatever and came around to the next bank and there was a cherry tree growing. I felt like I got my answer. That's happened a few different times, where I go back and see what I planted the year before sprouting and germinating in density. It's just a wonderful confirmation. And it's not all up to me. It's up to the space and the conditions. It has its own intelligence.

I visited these currant gardens way up in the middle of nowhere in Oregon and it was totally a huge garden. You could feel bear there, and see all the tunnels pushing through these big thick bushes. Very much planted, not by humans. Planted very much by bears. You could feel their energy there the whole time. I thanked them for letting me be there to collect some berries for pow-wow. It was absolutely planted by bears.

KOLLIBRI: By bears pooping you mean?

NIKKI: Yes. And who am I to say that wasn't intentional? Life still wants to happen even if things are changing. The nature of life is resilience.

KOLLIBRI: Is there anything else you want to say?

NIKKI: I think it's important that people give themselves permission to start participating with the world around them. I think that drastic changes have a tendency to make people feel shut down in different ways, but everything—the whole way that our society is living—is so destructive that we're afraid to do anything. But there are other ways. I feel like part of that process is giving yourself permission to interact and to explore what that feels like for you, see where it takes you.

28 July 2019

California Drivin'

California is the third largest US state by area and a tremendous amount of diversity exists within its borders in terms of plants, animals, climate, geology, politics, and more. I started traveling in California regularly in 2010, and since 2014 have spent more time there than not. Over this period, I grew to love the place, or more properly, the places. The cross-country trip I took in 2018 was a confirmation that no other US state holds the same attraction for me. Oregon is in second place but after that, whatever ranks third or greater falls much shorter in my affections.

THE CENTRAL VALLEY

The Central Valley of California is an agricultural wonder, producing about half of the fruits, nuts and vegetables grown in the United States. The soils, climate and availability of water provide conditions that are nearly ideal for farming. Or perhaps we should say, that "have been so far," but more about that later.

The Central Valley is divided into rough thirds with the Sacramento Valley in the north, the San Joaquin Valley in the middle and the Tulare Basin the south. It is bound by the Sierra Nevada mountains on the east and the Coast Ranges in the west. On all but the haziest or rainiest days, one or both of these continuous ridges is visible from Interstate 5, which traverses the entire valley north–south. Most of the valley floor is flat and low in elevation, low enough that next time all the polar ice is melted, it will be underwater; if any humans are around, perhaps they will refer to the Sacramento Sound and San Joaquin Straight and Tulare Bay.

As seen from the 5, the Central Valley is dominated by agriculture, and punctuated by just a few urban areas, the largest of which is Sacramento. Most are much smaller. Crops vary along the route, from olives in the north, especially around Corning ("Olive Capital of the US"), to almonds in the south, with grapes, hay, citrus, cotton, tomatoes and many other vegetables between. Everything is planted in large monocropped swathes, divided by ruler-straight lines. The rows of trees, flashing by at 70+ mph, are seemingly endless. When in flower in spring, the sight is quite the spectacle.

Rural roads cross over the 5 at regular intervals, and these elevated locations provide good vantage points. From such spots, one can't help but be impressed by the enormous scope of the Valley's plantings, seemingly filling the entire valley floor like a flood, and lapping up on the mountains along the edges. Satellite photography shows that there's nothing "seeming" about it. Virtually the entire Valley *is* drowned under farms.

Having taken in this sight, one should ask, "What was here before?" because all of these extensive operations displaced something. Many somethings. Besides plants and animals, entire ecosystems like wetlands, woodlands and grasslands were drained, cut, and scraped away. Wiped out were not only many species—trees, mammals, insects, etc.—but the aspects of place they co-created: habitats, relationships, and cycles.

To wit: The tree shelters the bird who feasts on the insects who pollinate the flowers who hold the stream bank which shades the pool where the returning fish lays eggs, dies, and is eaten by the bear whose scat fertilizes the tree with minerals from the deeps of the sea.

That's only one strand from a web of interactions complex beyond rational comprehension (which is not to say beyond human understanding) that was replaced, heartlessly, thoughtlessly, by rice, grapes, and alfalfa. Maintaining the domesticity requires regular treatments of mowing, plowing and spraying. We don't recognize this behavior for what it is: vicious.

The humans who were at home in the Central Valley before the European Invasion lived very well by all accounts. The unravaged landscapes were rich in food, including nuts, roots, berries, seeds and many kinds of game. Certain ecosystems were kept in high production with controlled burns, which mimicked natural events. These people did not live in one place all year round, instead moving between established camps depending on the season. Inter-tribal conflicts seem to have been rare. The Spanish missionaries were the first to intrude on this idyll with their crosses and cattle, but it was the 1849 Gold Rush—which brought 300,000 US Americans to the region—that spelled the end. Within a few short decades, most of the natural wonders of the area had been razed and plundered. This is the legacy of the farm fields in the Central Valley: the annihilation of thriving natural ecosystems and human cultures.

This won't last forever, of course, or anything approximating it. A relatively stable set of circumstances eased the transmutation of the wild to the domesticated in the Central Valley and allowed the new regime to thrive—such as that can be said—but conditions are different now. The soils are not as rich. The weather is not as stable. The water is not as plentiful. The peak has passed and the down slope threatens to be a rough ride. Those of us who wish for a restoration of the Oaks, the Antelope and the Monarch—and the Nomlaki, the Patwin and the Yokuts—will be disappointed if we think the Valley can go back to how it was. The climate is changing too much too fast for that. The former states of equilibrium are no longer possible. The approaching transitions will be tantamount to the impact of a large meteor, and could very well be nearly as abrupt.

If it's any consolation, the big rock that took out the dinosaurs was survived by the ground-dwelling ancestors of what are now the "most diverse and globally widespread group of terrestrial vertebrates"[1] on the planet: birds.

TWO-LANE DESERT BLACKTOP

There's nothing like desert driving.

I don't mean the freeway. Yes, the wide lanes and broad curves that let you cruise at 80+ without really paying attention are great for making good time, but they take you too far away from the landscape and from the act of driving itself.

I'm talking two lane desert roads; blacktop, but with unpaved shoulders, and sharper turns and steeper inclines than the freeway. Roads where you've got to watch for jackrabbits at dusk and you might even see a tortoise.

Tortoises at the side of the road always look grumpy. Biologists might say that's just the shape of their face, but I'm not so sure. It's not like they don't have anything to be grumpy about. Picture tortoise at roadside: *"What the <tortoise-curse> is this <tortoise-curse> thing in my way? It's too <tortoise-curse>ing hot, that's for sure. And what's that sound? No relation of mine, that's for sure. Here it comes and—what the <tortoise-curse>ing <tortoise-curse>? <Tortoise-curse>! That thing's gonna kill somebody! Wait, what's this? It stopped and opened its shell and now one of those cursed two-leggers is coming this way! Grrrr!"* Okay, it's not exactly like that; after all, tortoises don't growl.

Both danger and care are key ingredients of enjoyable desert driving. I know there's another level where it's all about four-wheel drive, high clearance and roll bars, but no. Kicking up dust and terrorizing the neighbors isn't my cup of tea. I'll stick to the asphalt, where tires belong.

Solitude. That's the other ingredient you never really get on the interstate. Even late at night, there's always somebody else. But going from Twenty Nine Palms to Amboy? Sure, you'll see *some* other cars, but not many. At the top of Sheephole Pass you can easily pull over and take in the amazing vista before you: below, a wide valley with an ancient lake bed (Bristol Lake), and beyond it, multiple ranges. Their colonized names are Bristol, Marble, Clipper, Granite and Providence. Regardless of what you call them, they are giants reclining in a colossal landscape and their magnificence is undeniable.

To the west is the Marine Corps Air Ground Combat Center, aka Twenty-Nine Palms, the largest Marine base in the world. What are the Marines doing so far inland? Blowing shit up, as can be heard in the nearby towns, including Joshua Tree. Over a hundred miles away as the crow flies is Camp Pendleton, a Marine base on the Pacific Ocean, above San Diego. Apparently, they practice beach landings there, then fly the personnel to to Twenty-Nine Palms by helicopter where they practice urban warfare in a fake town built there.

Big chunks of the landscape in southern California are controlled by various branches of the armed forces and since the "War on Terror" began, efforts to expand these chunks have stepped up. So, land that was publicly accessible is made into a bombing range, or the buffer zone around it, and now its inaccessible. People are split on whether this is good for the ecosystems or not. Obviously, it sucks for the Tortoise whose ancient home is now a playground for tanks, and moving them doesn't work well either; transplanted Tortoises die more than half the time. On the other hand, if the Tortoise—or the Chuckwalla lizard, Kangaroo Rat, Roadrunner, White-lined Sphinx moth, Painted Lady butterfly, Carpenter bee, Cholla cactus garden, Creosote Bush ring, *Cryptantha* patch, or an untouched fragment of cryptobiotic crust—is outside the active area but within the no-go zone, then they are now enjoying the protection of a *de facto* nature preserve, guarded by the military, where humans are excluded the majority of time. Not that I'm in favor. The US military budget should be cut by 95% and whatever personnel and equipment that remains repurposed to plant trees and help out during natural disasters (of which we'll surely be getting more). As for the Pentagon's vast real estate holdings, they should stay in public ownership but be made into *real* preserves. (Practically speaking, many ex-military areas will be dangerous to enter because of explosives, toxic chemicals, radioactive waste, etc., meaning that some setting aside will be mandatory.)

Sheephole Pass is at an elevation of 2307 feet and Amboy, 25 miles away by two-lane blacktop, is at 639. Watch your downhill speed—the California Highway Patrol lurks near the bottom of the hill and picks off people flying down.

An industrial salt evaporating operation squats in the Bristol Lake's dry bed, and the land around it is scarred. The first European settlers harvested salt from the surface, but contemporary methods use deep ditches to evaporate the underground brine.

North of the lake is Amboy, on the old Route 66. Amboy is not really a town, just a wide spot in the road with a gas station, post office, abandoned motel and a scattering of other random structures. Retro signage sets the mood. Tourists snap pictures of all of it. I'm not into Route 66 for its own sake—the natural environment fascinates me much more than the built one—though I'll admit to getting some kicks out of the period style. I can appreciate the space age aesthetic for its optimism, so much at odds with the dark moods currently favorable.

Amboy was a double boom town, first as a railroad station in the late 1800s, and again starting in the 1920s as a popular stop along Route 66. When Interstate 40 opened in 1973, bypassing Amboy, the town mostly dried up. Efforts to revive it since then have so far proven unsuccessful.

The Mojave Trails National Monument which surrounds Amboy on all sides, is named in part for Route 66, and in part for Mojave Road, an east-west route further north that was established by Native Americans. Spanish traders and missionaries used it when it was Mexican territory, and it became a wagon-trail route for settlers after the US Americans seized the area.

All the desert is stolen, of course, and the blacktop roads are vectors for trespassing. I get that. I don't really belong there. I can visit, but I shouldn't stay too long.

Where *do* I belong? In a certain sense, nowhere. I am a colonizer and my mother countries have no interest in accepting me back, not without a bigger bribe than I can offer. In a deeper sense, I could belong in any place that invites me, so it's a matter of being receptive to such voices. In yet another sense, home is simply the experience of living itself, when I am fully present with it. I venture to say that I've experienced flashes of that consciousness. Some have been inside, with candlelight and incense, listening to classical Indian ragas; others outside, when the evening air is full of frog song or the morning silence is broken by a certain bird call; still others, while driving by myself on a clear day in the desert on two-lane blacktop.

THE SALTON SEA

The Salton Sea is one of the most bizarre places in the United States. Located in the Imperial Valley of southern California, the body of water was created accidentally in 1905 during an attempt to improve the canals that brought water to the area from the Colorado River. A poor engineering decision led to a breach in the river's banks that rerouted most of the flow into the valley for two years, before the mistake was fixed. The result was the largest body of water in the state, almost twice the size of Lake Tahoe.

The Salton Sea was a popular tourist attraction until a deterioration in water quality led to dead fish and unpleasant odors. For an entertaining and informative account of the region's halcyon days, their decline—and of the eccentric characters drawn there—check out "Plagues & Pleasures on the Salton Sea," a documentary narrated by the singular John Waters. Something very uniquely "American" happened there for sure.

This is not the first instance that the Imperial Valley has hosted a lake in the same spot. Since at least the end of the last Ice Age, 12,000 years ago, the Colorado River has alternately bypassed the basin or flowed into it. Some of the previous filling incidents were recent enough that Native Americans remembered them and told stories to the early European explorers. In 1500, for example, there was Lake Cahuilla, which was 26 times the size of the current sea and left marks on the nearby hills like rings in a bathtub. Ancient fish traps that tribes made with stones can also be found on the nearby slopes. The last major in-flow happened from around 1700-1750, but by the time Spaniard Don Juan Bautista de Anza (for whom Anza-Borrego State Park is named) passed through the area in 1774, it had dried out again.[2]

The old pattern of filling and evaporating that occurred naturally for millennia has been interrupted in the current era by regular run-off from the irrigated farms around the sea, which keep the level from sinking too fast. However, this water supply is laden with fertilizers and the excess nitrogen and phosphorus have caused toxic algae blooms that have led to large scale die-offs among migrating birds. In 1992, 150,000 grebes died this way. In 1996, avian botulism caused by the poisonous conditions took out large numbers of White Pelicans (15-20% of the western population) and Brown Pelicans (at least 1000 individuals), who are endangered. 1999 brought the mass death of 7.6 million fish—tilapia and croakers—from lack of oxygen caused by algae blooms.[3] Evidence of this event is plain to see around the shores of the lake as a clearly demarcated band of skeletons. Toxins produced by some algae in the water are nerve poisons that pose great danger to humans (especially children) and pets.[4]

Water quality issues will worsen as lake levels sink, which they are guaranteed to do as the region's share of Colorado River water declines. The proportion of noxious substances, which includes sewage, will thereby increase. Additionally, shallower water heats up faster and to higher temperatures, triggering more algae blooms. As if all this wasn't enough, areas exposed by the receding shoreline will be a source of toxic dust. It's a disaster all around.

At first glance, the Imperial Valley doesn't look like an appropriate place for agriculture. The unirrigated landscape is desert, after all, and a vegetatively sparse one at that. However, the multiple floods over the centuries deposited generous amounts of silt in the valley, which provides a rich medium for crops. Just add water. (Which is exactly what people were trying to do when they accidentally made the sea.)

A variety of crops are grown, including date palms, sugar beets, sugar cane, bamboo, flax, and a variety of vegetables like carrots, kale, radicchio, lettuce, melons, broccoli, cauliflower, cabbage, potatoes, onions, sweet corn, bell peppers, chilies, artichokes, cilantro and salad mixes. The area also produces a major amount of hay, including alfalfa, for the dairy industry.[5, 6]

I have spent a fair amount of time checking out the agriculture in Imperial County around the Sea, both on my own and with Clarabelle. The style of farming there is intense. Big monocropped plantings leave no room for anything else in the landscape. Sprinklers are spaced with precision and deliver chemical additives as well as water. In some fields, acres of plastic are rolled out as weed barriers between rows. There is no sign of native vegetation. Insects are largely absent. So many plants yet so little life. The landscape was transformed from ecosystem to factory, and the only natural denizens are those who can fly in and out. (Birds do make themselves at home all around the area, especially where a field is flood-irrigated.)

Somehow, some of this mess is designated "organic." Clarabelle and I saw two trucks one afternoon, both loaded with carrots in open-topped bins covered with tarps, identical in all ways except that one vehicle was labeled "organic only." How different could the raising of these two harvests really have been? How "sustainable" is it to be growing carrots in the desert in the first place? Is this what the originators of organic farming had in mind? No, I can assure you, having met some of those very people in Oregon when I was farming there. Small-scale soil-building with a nurturing touch is what they were thinking of, not this mammoth, input-driven production machine. Real "organic" is far more than "no spray"; it is defined not just by what it lacks—poisons—but by what it boasts: a healthy biome, rich in minerals, shared with companions floristic and faunal, seasonally tuned and truly vital. Yes, this is rare, and of the items stamped "organic" at the store, only a small fraction of them are products of such circumstances. To further confound the situation, not every farm that is genuinely organic in these ways can legally use the label; the expense involved is too high for many. That's not an accident. The guidelines of the US Department of Agriculture's National Organic Program were defined in large part by Big Agriculture.

The only way to be assured that the food you're eating is actually clean, healthy, and raised responsibly is to know your farmer, of course.

But do I myself buy "organic" food from strangers or corporate entities on a regular basis? Yes, I do. Even though some of it is coming from places like Imperial County—which ought to be left to the birds—I still believe it's preferable to conventional, which is a total horror show. How many of our cancers, autoimmune diseases, and other ailments are due to the chemicals and genetically modified ingredients in our diets? We don't know, because we haven't done long term tests. That's what the population is being subjected to right now. So my choice to go organic is actually a contribution to science: someone needs to be the "control" group.

L.A. Freeways

Driving on the freeway through a big city you don't know can be exhilarating. At least for me, so I guess I shouldn't say "you." Plus, I'm about to talk about Los Angeles specifically, so "you" probably *really* don't want to be included.

I know I'm supposed to hate L.A., because apparently that's required if you're anyone from anywhere else: an East Coast intellectual, a Flyover Land redneck, a Pacific Northwest... well, *everyone* in the PNW hates L.A.. But *I* just don't hate L.A. In fact, I kind of love it. The sun, the palm trees, the speed: what's not to like?

For me, two aspects of L.A. comprise the initial impression. First is the buzz of being in a big city in general: the charge of so many souls packed in one place and the sensation of penetrating an energy field of that scale (and at a degree of undifferentiated consciousness exclusive of individual tensions, i.e., without egos). Secondly, there's the thrill of seeing the names of famous landmarks on exit signs: Beverly Hills (where you can find the famous zip code), Sunset Boulevard (where the paranoia ran deep in "Stop, Hey What's That Sound"), and Santa Monica Boulevard (where Sheryl Crow wanted to see the sun come up). It's a feeling of being *somewhere*.

But my favorite highway sign in L.A. isn't well known except to locals. Westbound on the 10, it announces: "Beach Cities." What an entrancing concept! Somehow both grand and charming, like something out of a children's book or a fantasy epic, where dreams could come true.

Which is totally a vibe that L.A. is imbued with, and no surprise. By sheer dint of the voluminous amount of media that's been produced there over the last century, the L.A. aesthetic has served as backdrop—seen or unseen, for real or at heart—in everything from soap operas to block-busters, porn vids to game shows, commercials to comedy specials. So when you walk onto the sound-stage yourself, or rather, race onto it along-side six other lanes, there's an undercurrent of cultural familiarity interjected with flashes of personal recognition—"Look, it's Hollywood Boulevard!"

There's no denying (much as many might like to) that L.A. is a promi-nent ingredient in the brand known as "America." It's just as central to that myth-making as anything that emanates from the corridors of power back east (either officially or clandestinely). The images displayed by L.A. on screens—sliver, flat or smart—connect with us subconsciously in part be-cause L.A. helped mold our subconscious.

Perhaps some sinister plot is being perpetrated. I don't know. But what I'm describing here exists on its own terms whether or not there are plans or puppet masters. There is "Action!" at a level where the kidder is also kidded.

An upshot of all of this is that neither hating nor loving L.A. matters to L.A., because nothing is important to L.A. other than its own projection of itself. L.A. embodies ego, and much ego embodies L.A. Unless you've been living in a cave your whole life, you've got some L.A. in you (and always will) and you haven't escaped from L.A. (and never will).

But back to the sun, the palm trees, and the speed.

L.A. taught me that the best way to drive on the freeway through a big city I don't know is to copy everyone around me. The posted speed limit is irrelevant. Go as fast or as slow as the locals. Don't race them and don't get in their way. If they want reasonable following distance, give it to them. (Doesn't apply in L.A.) If they like to tailgate, do the same. (Definitely ap-plies.) Individualism has no place on a crowded freeway; the daredevil and the slowpoke are both hazards.

Tunes are always a key component of my L.A. drives. I don't want to touch the phone while driving, so I either pick out a playlist long enough to last the whole day or put something on repeat. The first time I was be-hind the wheel solo in L.A., Jimi Hendrix did the trick; cacophony for ca-cophony. One afternoon on my way out of town, I listened to Beck. He's from L.A., so it was positively meta. Made me want to go up to Glendale with Debra and get a real good meal. Another time I programmed Stevie Wonder's "Maybe Your Baby" to loop and let it go for over three hours. That infectious funk—with its sultry Moog bass line, strutting stabs of clavinet, sinuous electric guitar, and hypnotizing ensemble of multi-tracked vocals (all of which are Stevie)—kept me alert but loose, in good spirits even when the going was slow.

Every time I drive in L.A., I'm super serious about focusing. The tangle of freeways is complex and forks can come up on either side. To make it easy, I will write down the highway numbers in large print with a sharpie in the simplest terms possible: "5 → 405 → 210 → 10" and navigate from that. I've tried to do the smartphone directions thing, but it just made me tense, and I actually got lost that way once (in New Jersey, not a fun place to get lost). For now I'm old school in that department.

So, pumped up with music and relieved of distractions, I can devote the bulk of my attention to the sport of driving in a crowd: knowing when to assert and when to give way, when to brake hard and when to just ease off, when to zip and when to drift. There's totally a flow, and I'm not going to call it zen, but I appreciate the awareness I must to cultivate to go with it.

The freeways in L.A. made me a better driver. You can be cruising along at 75 and then crest a rise to find a sea of red lights—brakes!—where traffic has become a parking lot. Such experiences also fostered an acceptance of whatever happens next. After all, nothing can be done to change what's ahead on the road. All that I (or you) can do is respond with sense.

And If my timing is just right, my view will be of palm trees in the sunset, zooming through Pasadena.

State Highway 20

For all the roadtripping I've done in California since 2010, there's only one highway that I've seen every mile of, and that's the 20, an east–west state route. A little over 225 miles long, it starts within sight of the Pacific Ocean at Fort Bragg, elevation 85 feet, and ends in the Sierra Nevada, at exit 161 of I-80, elevation ~5600, near the infamous Donner Pass. The entire length is driveable in a day—Google Maps gives an estimate of 4 hours and 45 minutes—but I've always taken more time myself.

In between are Redwoods, lakes, oak savanna, farm fields, wetlands, and coniferous forests. A dozen or so towns are arrayed along the route, all except one with less than 13,000 people. Though the population density is low, at no spot is the landscape unblemished by the touch of European-descended humans. As such, it provides a survey of exploitation, historical and contemporary.

If you wanted to explore the 20 starting in the west, a convenient place to spend the night before is Indian Springs campground, accessed from the next exit to the east, #164. The 28-site campground is next to the South Yuba River in the Tahoe National Forest. A large rocky outcropping just east of the campground is home to many native plants. When I visited in May 2016, I saw a sedums, wild buckwheats, ferns, mushrooms and Manzanita. An active railroad runs nearby. This campground is right next to I-

80, so it's never totally quiet, but the running water and verdant conifers make it a pleasant spot anyway.

The first leg of the drive, from I-80 to Nevada City/Grass Valley—a distance of about 25 miles—is a winding route through "Sierran Mixed Conifer" a habitat whose dominant tree species are White Fir, Douglas-Fir, Ponderosa Pine, Sugar Pine, Incense-Cedar, and California Black Oak.[7] Sugar Pines produce large, heavy cones nearly the size of a human skull. I would be shocked if any old growth trees still stand in the area at all. Logging was widespread here.

The road here is mostly downhill. Indeed, the highest point on the 20 is at or near the junction with I-80. Several to many campground and trailheads can be found along this section. In the winter, it's undoubtedly under snow much of the time, but in the summer offers a respite from the heat down below in the Valley.

This very rural area is followed by the most urbanized portion of Highway 20, where it's actually blown up into a full-fledged freeway for about four miles as it passes through Grass Valley. Grass Valley (pop. 12,860, elev. 2,411) is one of only three towns on the 20 where natural foods are available, and the town boasts not only a co-op but a couple of organic cafes.

But regardless of any "sustainable" businesses in the area now, it was settled during the Gold Rush for the purpose of exploitation. From 1850-1851, Nevada County was the leading gold county in the state. Further, Nevada City (pop. 3,068, elev. 2,477) was built on the site of Ustumah, a village of the Nisenan Native American tribe.[8] In the current day, the area is a hotspot for *Cannabis* cultivation due to its agreeable climate for that crop, its remoteness (despite legalization, much of the industry is still illicit), and its close proximity to the Sacramento market. The so-called Emerald Triangle counties on the coast have been losing pot business to this region.

After departing the Grass Valley/Nevada City conurbation, the 20 returns to two lanes and curves gently through Oak woodland interspersed with pasture as it descends into the Sacramento Valley.

Once on the Valley floor, Big Agriculture is king. Only a few trees break up an otherwise open, tamed landscape. Soon, Marysville (pop. 12,072, elev. 62) pops up, on the east side of the Feather River. Named after Mary Murphy Covillaud, a survivor of the Donner Party, it's another Gold Rush hot spot. The lure of wealth attracted immigrants from Asia, and like many other West Coast towns of the time, a sizable Chinese population lived there. But in 1886, also like many other West Coast towns, the European-descended population violently drove them out.[9] The anti-immigrant sentiments and policies of the present day, as bad as they have gotten, still don't match the sins of the past. But it is for those us alive now to prevent their repetition.

On the opposite side of the Feather River is Yuba City (pop. 64,925, elev. 59), the largest municipality on Highway 20. The word, "Yuba" might be a distortion of "uva," the Spanish name for grape, in reference to the wild grapes growing along watercourses in the area.[10]

Just west of Yuba City is a dramatic geographic feature known as Sutter's Buttes. The formation is known as the smallest mountain range in the world. Geologists do not agree on whether it is a stray piece of the Coast Ranges or, more intriguingly, the southernmost outlier of the Cascade Range. Though now named after John Sutter, at whose mill gold was famously discovered in 1848, it was originally called *Histum Yani* or *Esto Yamani* by the Maidu Native Americans and *Onolai-Tol* by the Wintun. These names are variations on "Middle Mountains,"[11] which describes them well.

The lowest point in elevation along Highway 20 lies somewhere between the Buttes and the town of Colusa (pop. 5,971, elev. 49 feet), I would guess where the Sacramento River crosses under the roadway. Colusa is a charming small town as measured by conventional aesthetics, with tree-lined streets of older homes, a classic main street, and the kind of well-proportioned architecture in public and commercial buildings that marked the early 20th Century. Of course, this was all cookie-cutter in its day, bore no relation to its local region, and thus can be fairly characterized as bland cultural imposition. In this way, it was no different than the homogeneous shopping strips that mar our cities now, other than being of higher quality construction. Back in those days, builders had access to grades of timber that barely exist anymore, and the fancy masonry work was affordable because the workforce was sweating it out without benefit of labor laws. Regardless of romanticized notions, that era is not when "America" was "great."

Between Yuba City and the junction with the I-5 are rice fields that are flooded regularly, much to the satisfaction of many migratory birds. These temporary water sources are a poor replacement for the millions of acres of wetlands that were drained, but they are "better than nothing" as people say. A network of National Wildlife Refuges also exists throughout the Central Valley, and one is located west of Colusa, on the south side of the 20. Clarabelle and I stopped there once and explored. Though completely surrounded by intensively farmed fields, the refuge definitely offered a haven. Besides seeing many birds, we found native plants, which are a rare find in the Central Valley.

Williams (pop. 93,670, elev. 82) is a classic cross-roads town. That is, it's primary reason for being is the junction of I-5 and the 20, which definitely counts for something. The I-5 is *the* major north-south route on the West

Coast, connecting San Diego, Los Angeles, the Bay Area, Portland and Seattle, so it's nothing to sniff at. As for the 20, if you're driving northbound, your next opportunity to head seaward isn't for another 65 miles, but that route, the 36, takes 143 miles and well over three hours to connect with the coast-following 101, as compared to the 20 here, which is merely 80 miles and an hour and a half. Furthermore, the next opportunity to cross over to the 101 from I-5 that *doesn't* take more time than the 20 is the 138 & 38 route in Oregon, from Sutherlin to Reedsport, but that's over 350 miles away. So if you want to get over to the west side of the Coast Ranges without taking half the day, this is your last chance for awhile.

Williams is one of three towns on the 20 with a prominent arch over its main street. The arch in Williams was nearly torn down in the early 1970's and was only saved by a citizen campaign led by local resident, Annette LaGrande. The 100th anniversary of its original completion was celebrated in 2018.[12] I've actually stayed the night in Williams on several occasions, either with Clarabelle or solo, because once upon a time there was a divey motel there for just $29/night. Unfortunately, they jacked their rates but didn't upgrade their amenities so it's no longer worth it, but during those previous visits, I got to know Williams a little better than most of the other nothing towns you breeze through. There's a good auto parts store, an honest mechanic, a decent Mexican place and Granzella's, a tourist attraction comprised of a family restaurant, deli, import store, bottle shop, cafe and sports bar with a motel next door if you need more than one day to take it all in.

West from Williams, the 20 takes you by Almond orchards and Rice fields. The mostly straight road has a couple curves of the "dead man's" variety, and I'm surprised they aren't marked with crosses and plastic flowers. Very soon you are climbing up into the Coast Range, leaving behind the Valley and its dismal agriculture. On clear days coming down this slope, you get a clear view of Sutter's Buttes and the snow-capped Sierra Nevada mountains beyond.

The first half of the way from here to the 101 is open Oak woodland and grasslands used by ranchers to graze cattle. The slopes and meadows are green for a brief moment in the spring, but most of the year are golden brown. The hills here are rounded and their sides smooth. It's beautiful countryside, even those areas that are filled with burned trees from the wildfires. The gatherings of blackened trunks are like crowds of statues all raising their arms together.

Prior to the Climate Crisis, wildfires in California were a perfectly normal event for such ecosystems, whose species are fire-tolerant or even fire-dependent. Native Americans set fires as part of their wildtending: Fir trees were prevented from over-running the Oaks; populations of acorn-predating weevils were kept in check; food plants that thrive in a post-burn environment were encouraged; and an open landscape benefiting the hunter was maintained. Also, roasted grasshoppers could be collected afterwards for eating!

But fire is not the same now. Higher temperatures and drier conditions are leading to events that are more intense, not sparing the older oaks as the smaller fires did. It's true that California's post-Ice Age climate has experienced much variation, including extended droughts, but the change underway now is heading out of that range. At some point—one that has possibly already arrived—conditions will have shifted enough that landscapes will no longer recover, and will transition to different mixes of vegetation. This is not a negative outcome in and of itself, but the emerging ecosystems will eventually not be supportive of humans or our endeavors.

As the highway gains elevation going east to west, it rises and falls, and features far more curves than straightaways. Pullouts are provided for slower traffic to allow others to pass, but more than half the drivers who ought to use them don't. Personally, I'm more than happy to get out of the way of people who want to get there faster than me, and I don't care if they're "speeding." I'm libertarian that way, believing that the best approach to driver interactions should be the collective effort to fulfill everyone's individual desires. Safety is obviously of paramount importance, but what constitutes that varies relative to driver skill, vehicle type, and road conditions, so it's realistic to be flexible.

All that being said, I must complain about one driving habit common in California, and that's tailgating, especially at high speed. I'll be cruising along a two-laner, having a lovely day at my own rate, when up behind me zooms Speedy McSpeederson who proceeds to sit right on my ass. On one hand, I get that a clear message is being delivered: "Get out of the way," and I can respect that, and sometimes when I comply at the next pullout, they'll give me a friendly double-tap on the horn as they pass. On the other hand, the proximity can be nerve racking as hell, especially when there's nowhere to move over or if I'm behind a bunch of traffic myself. I've also had days when tailgaters haven't perturbed me in the slightest; when I've recognized that their agitation is entirely their own and doesn't involve me at all. But much as some people might make claims to the contrary, there's no switch to flip for zen, so sometimes, these jerks just annoy the hell out of me.

The next section of the 20 passes through a string of villages along the north side of Clear Lake: Clear Lake Oaks (pop. 2,359, elev. 1335), Glenhaven (pop. 325, elev. 1345), Lucerne (pop. 3,067, elev. 1328), Nice (pop. 2,731, elev. 1362), and Upper Lake (pop. 1,052, elev. 1342). On the northbound 101, the exit for the eastbound 20 advertises "Lake County Resorts" and that's some of what you'll find along here: motels, RV campgrounds, and other vacationer spots. Only one of them is upscale; otherwise, this area has an unpretentious, down-home atmosphere. Unlike the towns back in the Valley and in the Sierra foothills, these are not Gold Rush settlements, and date from the teens and twenties. Highway 20 did not open as a fully paved route until 1932, and the last portion completed was in this lakeside section.[13]

Clear Lake happens to be the oldest lake in North America, at ~480,000 years old. As explained by California writer, Joe Kukura, most lakes disappear within 10,000 years because they fill with sediment. But beneath Clear Lake is molten rock that "tilts" the lake bed every year, so that sediments "seep into the gaps" and don't collect.[14] This charming fact is offset by the lake's pollution. The Sulphur Bank Mine near Clearlake Oaks, which operated for a century starting in 1856, is responsible for heavily contaminating the lake with mercury. The old mine is now a Superfund site, and cleanup activities, already having cost tens of millions of dollars, are incomplete and ongoing.[15] The state of California advises children and pregnant women to limit their consumption of fish from the lake.

Walnut orchards are common along the 20, from Lucerne west. Parts of that town were build in old orchards, so many houses have Walnut trees in their lots. I have spent some gathering Walnuts there and in other parts of the county. North of Upper Lake, for instance, Walnut orchards are the dominant form of agriculture. Upper Lake, by the way, is the second of the three towns on the 20 with an arch over its main street, and is worth visiting if you want a good coffee. The only other decent joe between I-5 and the 101 is a place in Clearlake Oaks, currently called Buddy's, but which I first appreciated under a previous name.

The remainder of the drive to the 101 is a mix of flat land with agriculture and hills with Oaks. Other landmarks include a Native American casinos, Blue Lakes, a berry farm featuring "Ollalies" (a Native American word for huckleberries), more scorched Oaks, and Lake Mendocino. The 20 then merges with the 101 near Calpella and heads north to Willits.

Willits (pop. 4,888, elev. 1391) would be more appropriately discussed in a piece about the 101, which bisects it entirely north-south (or did, until the bypass was completed in 2016) and since the 20 turns off for the coast partway through town, before reaching Willits' original downtown section. A few of the town's amenities are worth mentioning, however. If headed east, this is your last chance to visit a natural food store until you get to Yuba City, so if you're planning to hang out in the Coast Ranges for awhile, you want to stock up here. Willits also has gluten-free pizza, a decent library, more than one good Toyota mechanic(!), a great used book store, a first-run movie theater, and an an arch over the main street (#3 or 3), which was a gift from the city of Reno and was localized upon installation with the declaration for northbound drivers: "Gateway to the Redwoods."

That this claim is true is sad, since Willits is in the middle of the northern part of the Redwoods historic range, which extends all the way down to Big Sur. But with over 95% of the mighty trees razed, and the bulk of the remainders found in Humboldt County to the north, it is remorsefully accurate to call Willits the "gateway."

All in all, Willits is a likable hamlet so it's unfortunate to hear that since the bypass opened, businesses in town have been suffering. The closing of the Paradise juice bar and espresso shop, for example, was a real loss. Price crashes in the *Cannabis* market have also contributed to the downturn.

I've only been to Fort Bragg (pop. 7,273, elev. 85) twice, so I don't know this stretch of the 20 as well as others. The first time was with Clarabelle, when we drove out to the ocean seeking relief from wildfire smoke in August 2017. The second occasion was to the Humane Society in the Spring of 2018 to drop off "Hightops," a stray cat who showed up at the cabin where I was staying in the hills outside of Willits. I had been feeding her for a couple months, but was leaving for an indeterminate amount of time and there was no one else around who would see to her care. It was a sad parting, and I still think about her sometimes. I don't worry though; being as friendly as she was, I'm sure she endeared herself to someone else.

The drop in elevation from Willits to Fort Bragg is over 1300 feet and the curvy distance is a little under 35 miles. A few stretches are quite steep. The territory is rugged and heavily treed, although almost exclusively by second growth, with virtually everything close to any road in this part of the world having been cut long, long ago. Nonetheless, there are some beautiful scenic vistas and a few places to camp. My friend and I found lots of mushrooms on our visit, though none edible.

Technically, Highway 20 terminates when it hits Highway 1, just south of Fort Bragg. However, if you go just a little ways north and take a left, you are delivered to a low rock wall with a big yellow sign declaring, "END" (as seen on the back cover of this book). For me, that spot is the real conclusion of the 225 mile journey because it is only there—after witnessing "managed" forests, worked-over fields, ranched savannah, mined hills and tainted lakes—that one finally gets to look upon something untamed: the ocean.

July 1, 2019

NOTES:

[1] Everything Dinosaur. "How Birds Survived the Cretaceous Mass Extinction Event." https://blog.everythingdinosaur.co.uk/blog/_archives/2018/05/25/how-birds-survived-the-cretaceous-mass-extinction-event.html [retrieved 7/8/19]

[2] The Salton Sea Authority. "Timeline of Salton Sea History." http://saltonseaauthority.org/get-informed/history/ [retrieved 7/8/19]

[3] *ibid.*

[4] Wilson, Janet. "Sunny days, fertilizer runoff and sewage can create toxic bacteria at Salton Sea" Palm Springs *Desert Sun*, April 4, 2019.

[5] University of California Cooperative Extension. "Imperial County Agriculture." https://vric.ucdavis.edu/virtual_tour/imp.htm [retrieved 7/8/19]

[6] Imperial County Farm Board. "Quick Facts About Imperial County Agriculture." https://www.co.imperial.ca.us/AirPollution/forms%20&%20documents/AGRICULTURE/QuickFactsAboutIVag.pdf [retrieved 7/8/19]

[7] Allen, Barbara H. "Sierran Mixed Conifer" in "California Wildlife Habitat Relationships System," California Department of Fish and Game, California Interagency Wildlife Task Group.

[8] Wikipedia. "Nevada City."

[9] Wikipedia. "Marysville."

[10] Wikipedia. "Yuba City."

[11] Wikipedia. "Sutter's Buttes."

[12] Green Jr., Lloyd. "Pioneer Day celebrates 100 years of the Williams arch." Williams *Pioneer Review*, June 26, 2018

[13] Wikipedia. "California State Route 20"

[14] Kukura, Joe. "Meet the Oldest Lake in North America." California's Lake County blog, June 22, 2015. https://lakecounty.com/blog/clear-lake-the-oldest-lake-in-north-america/ [retrieved 7/8/19]

[15] Wikipedia. "Sulphur Bank Mine."

What to Do at the End of the World?— A Talk with Kevin Hester

Kevin Hester is an environmental and anti-imperialist activist living in New Zealand who is raising the alarm about the dramatic, planetary-scale changes that are underway. He is expecting "the imminent collapse of the biosphere from the perfect storm of runaway abrupt Climate Change and indifferent human hubris." He regularly publicizes pertinent stories and data on his website, kevinhester.live, and on his interview show, Nature Bats Last (on the Progressive Radio Network) which he hosts with Guy McPherson, the well-known proponent of Near Term Human Extinction theory. In January 2018, I spoke with Hester over Skype. What follows is a transcript of the conversation, edited for clarity.

SONNENBLUME: So I don't know very much about you except that you're an activist around Climate Change and specifically around abrupt Climate Change.

HESTER: Correct. And also, I've been involved in anti-nuclear, anti-racist organizations all my adult life as well. So I've been in that whole spectrum of environmentalism and geopolitics, the whole gamut.

SONNENBLUME: New Zealand doesn't have any nuclear there, right?

HESTER: Yes. Myself and my peers played a small role in that where we used to protest out on the water where the nuclear ships used to come to New Zealand. We're a member of the ANZUS treaty (Australia-New Zealand-United States) and we fought the government into a corner and backed them up and got enough public support to have New Zealand declared nuclear-free. In the very early 1990's. It was a really big victory for us.

SONNENBLUME: And that's still the case, right?

HESTER: It is, but incrementally the government of New Zealand is making a rapprochement with the Americans. I would be certain that

American ships are using our international waters for ships with nuclear weapons, and probably nuclear-powered. The government just turns a blind eye to everything.

SONNENBLUME: Right. It seems like that's kind of the story these days.

HESTER: Yes. Subservience to the empire. Whatever America wants, America gets. They get it either officially or unofficially, from what I've seen.

SONNENBLUME: There's very little awareness of the United State's imperial status within the United States.

HESTER: Yes, it's extraordinary for us anti-imperialists on the outside looking in. It's incredible. But one of the points I make all the time about imperialism is that you can't be an environmentalist and not be anti-imperialist because the US Pentagon is the single largest consumer of fossil fuels on the planet, so the carbon footprint of the US military machine is enormous. And then there's the massive carbon footprint of all the nations responding to it in kind and arming up. So the warmongering of the planet is probably the single individual contribution to Climate Change.

SONNENBLUME: Yes, certainly. Plus, the fact that the United States behaving as it does prevents cooperation from occurring.

HESTER: Yes, that's right. It's the behavior of the bully. So people are bullied into submission, into doing what the United States wants. or else they're ostracized and destabilized, and regime changed. That's what we seem to be watching everywhere these days. It's a pretty bad, disgusting way to be watching the human story on this plant coming to an end.

SONNENBLUME: I've been an activist for a number of years, too. Not as long as you have, and part of what I've been curious about is tracing back, where did the trouble start? In part to know if there's any way to get out of it or, if there's not any way of getting out of it, at least what's a better way to behave as we go down, you know? That's what has led me, for the time being, to agriculture. The imperial project started with agriculture and with urbanization.

HESTER: Yes. When civilization took off, [with] the ability to store grains in scale. It was also the cul-de-sac that has taken us close to extinction because we just raped and pillaged everything that we saw in our way. We've lost so much soil and soil quality in the last few decades. Like a lot of Climate Change graphs, it's exponential.

SONNENBLUME: Yeah. They say that, for example, in Iowa, in the United States, that two bushels of soil is lost for every bushel of corn that is harvested there.

HESTER: Yeah, when you see the statistics like that, you realize what an extraordinary level of overshoot we are in, with the human population and industrial civilization and the over-consumption. It's going beyond our planetary boundaries. It's no surprise that we're hitting the wall. But it is surprising that people have been talking about it for decades and, now that we're hitting the wall, they literally can't accept it. It's quite amazing watching the cognitive dissonance and the denial. And the behavior of people who should no better. They just can't accept the fact that we've hit the wall. It's quite an extraordinary dichotomy.

SONNENBLUME: When you say "hit the wall," I think I know what you mean from having read things you've put out there, but could you say what you mean about that, in a real basic way?

HESTER: Yeah sure. I believe that collapse of the biosphere is underway. It's not going to happen sometime in the future. It's already underway. I interviewed Professor Paul Ehrlich for my radio show Nature Bats Last on the Progressive Radio Network and one of the papers that Paul and his colleagues have just released was called "The Unfolding Sixth Great Extinction and Biological Annihilation on the Planet." What it stemmed from was that just recently they discovered in Germany that they've lost 80% of the flying insect population in the last 17 years. That's in a reserve system in the normal countryside, not where agriculture is bathing the land in herbicides and pesticides. This is in the national parks [where] they've had this collapse. It's mimicked the collapse of the phytoplankton and krill species which are keystone members of the marine food web. So it doesn't take a rocket scientist to look around and see that collapse is unfolding.

SONNENBLUME: Yeah, the insects; that story was released this last year, I remember, by those German scientists and I think there's been some supposition that part of that was due to the use of agricultural chemicals and conventional farming. But it's beyond just that I think.

HESTER: Undoubtedly. It's habitat loss. Habitat is collapsing on the planet. It's collapsing for these keystone species. People just don't understand how important these minor—physically minor or small insects and organisms have toward being key links in the chain of the biosphere. We've been chipping away, chipping away, chipping away and it's incredible what we don't know. Most of the species on this planet have yet to be identified. We've finally just barely had a quick look at the bottom of the oceans, which is where we all come from. We're losing keystone species that we know of. But we're also losing keystone species that we didn't know existed.

SONNENBLUME: Right. The phytoplankton in the ocean, that's considered a keystone species because they're at, quote, "the bottom" of the food chain.

HESTER: Yes and they do multiple roles. They metabolize carbon. Most of the carbon that we emit goes into the oceans, not the atmosphere, and until recently a lot of that carbon has been metabolized by the different species like the phytoplankton. And the carbon has been stored at the bottom of the ocean and oxygen has been released. But what we're finding is that we're now getting a situation where the oceans, which effectively are like a battery, are overcharged and acidifying and there's evidence that the ocean is out-gassing carbon now. So we're losing our sinks, our sinks that have buffered us from the worst of the damage. Those sinks are now falling over. Last year we lost more forests globally than the physical size of the country that I live in. New Zealand is actually a very big country. People mistakenly think that it's a little country because there's only four million of us but the actual physical size of the country is very very large. [According to Wikipedia, the area of New Zealand is 103,360 sq miles / 267,710 sq km, which is "slightly smaller than Italy and Japan and a little larger than the United Kingdom."] And we've lost that much forest off the planet. So that means that all that carbon, when it burns, has been released back into the atmosphere and then the carbon sink component of those trees is lost so they're no longer absorbing the carbon. This is the perfect storm in exponential function.

SONNENBLUME: And the forests are being cut down in large part for farming, it's my understanding, for raising cattle, for raising feed for cattle, and also for palm oils.

HESTER: Yeah, all those things are happening, everywhere you look. It's extraordinary. And of course as our atmosphere heats up, trees stop sequestering carbon, somewhere around 35 or 37 degrees [Celsius, which is 95-98.6°F]. Once they get into drought and heat stress they become carbon emitters. We had a situation a few years ago where they had thousands of fires burning in Indonesia and for a brief period of time, for a few weeks, Indonesia was the second largest emitter of carbon on the planet.

SONNENBLUME: Just because they were having fires?

HESTER: Yeah. And a lot of those fires were capitalist fires where industrialists had paid people to go set fire to native forests. Because they were in drought, and it was so hot, they could get them to burn. Because traditionally a lot of these places were rain forests. The rain forest wouldn't be able to burn because it was green and damp. But when it gets into heat stress, then these lunatics can set fires and burn it and later on they can go back and plant palm forests on them. But another thing that takes place when land is burned off and—this happened in California just recently [when] they had those massive fires—and then afterwards they had deluges and those deluges washed an extraordinary amount of topsoil into the rivers and oceans. Lost! Completely lost.

SONNENBLUME: I was following the story of those fires. You're talking about the Thomas Fire that just happened around Santa Barbara. This summer I was working in northern California doing some farm work for some friends when those fires were happening and there were days when I needed to be wearing a mask working outside. You didn't want to exert yourself without wearing a mask because there was that much smoke in the air. At the same time, the nature of agricultural work is "this needs to happen now, you need to do that now." So—

HESTER: You don't have any choice. Of course. And a lot of agricultural workers are low income workers, and they desperately need every day's wages, so it's easy for exploitative employers to get people to go and work in those dangerous conditions because they desperately need the day's wages. This is the whole thing about the reserve army and capitalism exploiting working class people. They get them to do shit that no one else would do.

SONNENBLUME: There was a photograph that came out during the Thomas Fire that showed the billowing smoke in the background and then in the foreground there were some agricultural fields and there were some Latino workers who were out there doing the harvesting. And of course it wasn't white people because it's much easier to exploit the migrants than the citizens here.

HESTER: Yeah same old story. Disenfranchised. Take advantage of the poor. Classic, quintessential capitalism where they exploit all the resources and they treat the humans as resources, they treat the flora and fauna as resources. Everything is just a resource to be exploited for capitalism.

SONNENBLUME: I agree, describing this as capitalism is perfectly accurate. And yet, capitalism itself doesn't go back six or eight thousand years, during which all of this destruction has been happening. I mean, capitalism ramped it up to a whole 'nother level. But—

HESTER: I think part of it was civilization, where we became civilized and started living in cities. One of the things with cities is that very quickly they over-reach their ability to feed themselves, and their supply chains get longer and longer and longer and longer. With civilization, where you can store grains, you can control people. In the days gone by when people were only hunter-gatherers and tending small plots, everyone in the village had to play their part or they got drop-kicked out of the village, but now it's different where it's all about exploiting. Oxfam just released a paper about wealth disparity on the planet. It just goes to show that literally a handful of people own half the planet's wealth. Wealth disparity is one of the driving issues of the sixth great extinction.

SONNENBLUME: It's amazing to think what a small number of individuals it is. I think I just saw that there are 8 individuals who have as much wealth as the, quote, "bottom" three and a half *billion*—with a "b" as in "boy"—people in the world.

HESTER: Correct. That's the stat I've seen as well.

SONNENBLUME: Well that's insane. Part of me looks at that and is like, how does that keep going? That's only eight people "with names and addresses," to quote Utah Phillips. So how does that keep going?

HESTER: The theory I believe is that capitalism is a vortex of socio-pathological people at the top. If you look at the people who are running the big corporations, that are grinding the living planet into dust, I believe that the vast majority of those people are suffering from psychological illnesses. You'll hear people say all the time, "Well how can they sleep?" making those decisions. It's a mistake to think those people have empathy. They don't have the same problems that we have. They don't have sleepless nights worrying about poor people down the road who are exposed to the elements because capitalism has just consumed everything around them and it's all been vortexed into the hands of the few. Those few don't lose any sleep over the poor. They don't lose any sleep over deforestation.

SONNENBLUME: I agree with you completely about that. I think it's a disease of the mind and perhaps deeper than that. I think it goes just beyond those few number of owners, too. Returning to what I said earlier about, here in the United States, very few people know that it's even an empire, I feel like I see this ability to sleep—regardless of what this culture is doing —is everywhere. It's kind of the normal behavior here. There's only a small number of people here who have empathy.

HESTER: Yeah, I think that has a lot to do with the fact that your media has been so conquered by the corporations. The same people who own everything on the New York Stock Exchange own the media that tells the people what to believe and what to think. There's been a little bit of a revolution with social media where anyone who's motivated enough can find critical examination of what's going on, and you can distill it and decide for yourself what you believe is happening on the planet. The average Joe, from what I can see in the United States, is living on extraordinarily poor nutrition. The food that they're eating, and then the nutritional value of the media they're getting, is zero is well. So these big corporations like Fox and CNN and all the big ones, they're just feeding people hogwash.

SONNENBLUME: They definitely are. I've been a media activist, so I've noticed these things about the media concentration—the concentration of ownership. And I feel that at this point, with the recent attacks we're seeing on alternative media by the social media giants, that the golden age of alternative media, at least in the US, has passed. We're past the peak at this point. Now we're in a time of decline, at least in the United States, for the power that alternative media can have.

HESTER: Unquestionably. I think we have to realize that the nefarious bastards that are behind all this have been working on this project on how to reign in the internet and they've been working on it a long time. I think Internet freedom is really under attack from many different respects. In the United States, as an outsider looking in, it looks like a police state to me. The state can spy on anyone it wants to. The beautifully written Constitution that you have has now been used as toilet paper and I just see the rights of the individual and media being eroded on a day to day basis. It's extraordinary.

SONNENBLUME: That's exactly what's going on here. I've personally spent more and more of my time in rural areas and camping on our public lands in order to get away from a lot of that and have some peace of mind and to be able to reflect on the whole situation and understand it better. Because the one thing that's been holding me in the United States for the moment—because at some point I might want to go some other places— has been the beautiful wilderness that still exists in the United States, especially in the western half. There's mountains and deserts and forests. And although you can't go anyplace that's untouched, you can go to places that have only been lightly touched. You can go to places in the desert where you can actually just be by yourself and not talk to anyone for weeks if you want. The perspective that I've been able to get from having time like that has really helped me personally. To be away from it all.

HESTER: I think it's an antidote for despair. For anyone who's paying any attention to the unraveling of the biosphere, the biggest challenge we're all going to have going forward is dealing with our grief. I believe that everyone is grieving to some degree or other, whether its consciously or unconsciously, and one of the ways to deal with that is to immerse yourself as much as you can in the natural world. It's easy for me because I come from an underpopulated, beautiful country surrounded by oceans so the worst horrors of Climate Change are still delayed here because of the effect of the cooling aspect of a) the oceans that surround us and b) we're quite low in the southern hemisphere. But what I always say, and what I do, is I spend as much time as I can in the natural world while it's still holding on. I'm bearing witness to the unraveling. It can be very alienating. You know, no one wants to hear me talk about collapse; they think I'm dire and doomy. But this is one of the problems with liberals, with middle class liberals: their lives are so comfortable, they can just ignore the fact that we're losing 150-200 species every day and the poor everywhere on the planet are just being crushed.

SONNENBLUME: Yeah. The effects of Climate Change are being directly felt by other people. I have a friend in Portland who was working with the Climate Change group, 350, which you've probably heard of. The local chapter of 350 in Portland was more radical than the national organization. They had a visitor, a Climate Change activist from Bangladesh, who came and lived at my friend's apartment for a few months and who was working with 350 PDX. And in Bangladesh there is no such thing, really, as Climate Change denial because they're all experiencing it there. Most of their nation is only a few feet above sea level. And what was happening there, that she reported, is that the freshwater is getting inundated with salt from the ocean and so people are getting sick and dying sooner than they normally would have, so women are now getting married and having children earlier than they used to because they can't count on being as old as they used to be when they had kids.

HESTER: Extraordinary. There's a parallel with that happening to my Polynesian and Melanesian neighbors. The wells and the aquifers on a lot of the island in the south Pacific are getting more and more brackish. A few millimeters of sea level rise can alter the pH of the land and the wells and aquifers on these islands. Most of the sea level rise so far has come from thermal expansion, where the oceans have expanded, not from melting ice caps and glaciers.

SONNENBLUME: Just from the fact that the oceans are warmer?

HESTER: There's a thermal expansion in the ocean. See, you heat water up and it physically expands in size. So that's where a lot of sea level rise has come from already.

SONNENBLUME: So you know people on some of these islands?

HESTER: Oh yeah, totally. I've got a very dear friend who is a professor at Victoria University, Pala Molisa, who comes from Vanuatu. I attended a conference that Pala organized at Victoria in Wellington, which is the capital. And there were a lot of people speaking there, a lot of indigenous people from around the Pacific. And one criticism of me when I speak, sometimes, is people say, "You're too emotional," and I find that hilarious that the only people who've ever said I'm too emotional are the Anglo-Saxons. None of the indigenous people have ever said to me, "Oh you're too emotional," because they're going through this emotional roller coaster. They know— they know that their islands are becoming daily more inhospitable to life. And the critical thing for a lot of indigenous people is that they believe that their ancestors walk amongst them. So it's incredibly important for

168

them to remain in their ancestral homelands because multiple generations walk amongst them. So when you talk to people in Polynesia about having to evacuate from Kiribati or Tuvalu, people don't understand what a cultural upheaval this is for these people. It's not just about leaving home. It's about abandoning your ancestors. This is why the world is dominated by the ex-Anglo-Saxon colonialists. There's so many of these important things that just go straight over their heads.

SONNENBLUME: One of the people that I interviewed for a book [I wrote] is a Cherokee man who is also a college professor of Divinity at a university in Portland.* He talked about how one of the differences between the indigenous worldview and the western worldview is that the western worldview is based in time and the indigenous worldview is based in space or locality.

HESTER: Very interesting observation!

SONNENBLUME: Yeah, and I would never have come up with that on my own. That's why it was so amazing to talk to him because he had these insights. That speaks exactly to what you were just saying about your Polynesian friends.

HESTER: When we were children, my dad used to mock ownership of land. I only would have been ten or something like that [and] I remember him pointing at a rock and saying, "Son, look at that rock. It's been here millions, perhaps billions of years, and then someone who looks like you and me walks by and says, 'I own it.'" He could just not take land ownership seriously. My dad was an advocate of custodianship, or being a guardian, and the idea was that if you were a guardian, then you had responsibilities toward [the land]; you need to leave it in as good of condition when you leave as when you arrived. And if we did that, we would have a habitable planet. But of course what we do is we exploit and we use. Our whole capitalist system is derived around an infinite growth paradigm and it's blatantly obvious that you cannot have infinite growth on a finite planet. Sooner or later, you're gonna hit the wall. Edward Abbey, the desert anarchist, has a famous quote saying that, "growth [for the sake of growth] is the ideology of the cancer cell." The planet's got cancer and it's really hard and bitter to accept that I'm the species that is the cancer. And then I'm a white male. I'm the worst. The most damage is being done by people who look exactly like me. That's part of the reason I spend my life on Climate Change awareness and these issues; because I'm the personification. If you look in the mirror, I'm the personification of evil.

---

* *Randy Woodley is his name. See the chapter, "Decolonizing the Western Worldview," in my book, "The Failures of Farming and the Necessity of Wildtending" (Macska Moksha Press, 2018).*

SONNENBLUME: Right. I understand that one. My background is Norwegian and German. I'm "Aryan" as they would say. We have a responsibility to use our privilege to try to do something good in the world.

HESTER: Yeah I hundred percent believe that. Through my activism, when I was in my teens, I was involved in anti-nuclear and anti-Apartheid. I was involved in an organization called HART which originally stood for "Halt All Racist Tours." New Zealand had a big connection with South Africa [with] sending rugby players over. The Apartheid state was so racist to its people and we fought that battle. There was a famous case [in 1981] where the South Africans came to New Zealand. They were called the Springboks —that was the name of their team. They were playing at a provincial town called Hamilton and there was huge anti-Apartheid riots outside. They ripped the fence system out. They occupied the field. And this was the first game that had ever been live-broadcast in the Apartheid state so everyone in South Africa was listening to their rugby team playing the world famous All Blacks [the name of the New Zealand team]. The ground was invaded by protesters and the game was stopped. Nelson Mandela wrote a letter to HARP from his prison cell on Robert Island and the broad detail of the letter was: "You have shone a ray of sunshine into a very dark cell." How's that for a trophy for everyone!

SONNENBLUME: Wow. That's amazing. I don't know if I'll ever accomplish anything as big as that in my life!

HESTER: We didn't know we were doing big things. This is the motivation, that you should just do what's right [and] don't be too linked to the outcome. I'm fighting for Gaia and I'm fighting for the natural world and I believe the biosphere is in collapse. I believe that it will unravel in an exponential way in the coming months, or years if we're lucky. If we get years we'll be lucky. But I still believe what you've gotta do is do the right thing until the last moment. I work on a little island where I've got a lovely couple of dear friends of mine who have set up a not-for-profit nursery where we're propagating native trees. We've got a rewilding program for our island. So any of the private landowners who come on board, they can buy the native plants from us at cost price, already on the island and acclimatized to the island. We're trying to rewild and re-vegetate. And I firmly believe that the planet will become uninhabitable and inhospitable for complex life very soon. But I would love to be planting another tree on my last day when the towering inferno sweeps across us.

SONNENBLUME: I understand. I feel the same drive.

HESTER: It's the only moral position to take. And you know, you talked earlier about 350 and how that one chapter was more radical than 350 nationally, and that's a really big problem for all the organizations like that, for 350, for Greenpeace. The activists have to put the acid on the corporate structures that are now running those organizations. None of them are facing up to the severity of the crisis. None of them are speaking truth to power. And it's up to all the good activists to take it on the chin and be critical about our own organizations. There's too much going along with historical allegiances. I've been working alongside Greenpeace for nearly forty years and we've done a lot of good things together, but the corporate structure now is our enemy. It's so much about revenue, bloody petitions. I've been petitioned to death. The planet's been petitioned to death. We need to stop things. We need to turn off the valves. We need to stop the trains and the ships in their tracks, not just petition pathetic organizations to listen to us. It's not having any effect. The old thing about going out to the road and shaking our banners, that's all well and good, but we've got to stop ships and we've got to stop the trains and we've got to stop the chemical pollution at the gate. We've got to take on capitalism and the corporations where it hurts the most, which is the balance sheet.

SONNENBLUME: Right I totally agree with that and I think that in the United States some well-meaning people get distracted by the, "Well, let me change light bulbs in my house" level of choices.

HESTER: Yeah and those things are important to do. But it's important to remember in the bigger scheme of things what effect that will have. And they will effectively have no measurable effect. If you want to do something really constructive for Gaia, it would be to stop the fighter jets and the warships. Keep the fighter jets and the drones on the ground and the warships in port. That's a meaningful thing that we can do. We do all these things: we try alternative energy—I have a solar powered house on the island, I catch my water off the roof, we grow our own food out here, we try to do whatever we can—but you've got to be realistic. It's not going to change anything. It's not going to solve it. What it does is make a privileged few feel better about themselves for their excessive lifestyles.

SONNENBLUME: Right. I mean it's the same way that it doesn't make a difference if you yourself have a vegan diet, in terms of stopping industrialized animal slaughter. But to not eat that food is still the right thing to do.

HESTER: Oh absolutely! Absolutely. Some people who believe what I believe—that we're already in collapse—are choosing to either not do anything or [are] discouraging people from doing anything. And that is completely a misrepresentation of our position. We're just being honest about the severity of the crisis. I am going to keep doing the right thing until the lights go out. All the really true environmentalists and people with their hearts in the right place will do that no matter what. But let's just stop bullshitting our youth. It really pisses me off when I see all this pressure put on young people that "you've gotta fix" this complete catastrophe that we have bequeathed them. That is so unfair. It's wrong to tell the young people that they have to fix the unfixable. All we have to do is be completely honest with them and say we've dumped this on you and the place is becoming uninhabitable really quickly. We're really sorry. You people have got to do the right thing for you. What I would suggest that young people [do] is learn horticulture, agriculture, how to fix things, now to make things. Learn practical skills. Don't go into law and accounting and IT. All of these things depend on the internet working. The internet 's going to go off soon. The more practical skills you have on that day, the better off you will be.

SONNENBLUME: I feel like the internet will not last forever and I've never treated it that way, because I remember, of course, before there was one. I worked in that field for awhile during the Dot-Com boom, all the while knowing this is temporary. This is a temporary little blip here that we have this global interconnectivity and if something's really important to me, well I'm not going to store it in "the cloud," I'm going to have it on paper.

HESTER: As part of my studies, I've been studying the fragility of complex societies. For anyone listening to this, I would suggest that they go to Youtube and search, "Joseph Tainter," and look at the different presentations he's done over the years about how the more a complex a society becomes, the more fragile it is and easy to knock over. I'll give you an example. Two years ago we had catastrophic flooding in Thailand and Vietnam. Those floods in Vietnam and Thailand shut down the Japanese car industry for a week because there was a certain couple of solenoids that were only built in one factory in Thailand and when that factory got flooded out, none of those cars on the production line could go through because they didn't have a vital ingredient.

That's what the nature of complex societies is like. That's why they're so fragile and certain things can come out of nowhere—you know, black swan events—can come out nowhere and tip the whole thing over. A little known detail about the degree of the predicament that we're in is a thing called "global dimming." If you look at organizations like Deep Green Resistance, they've been advocating for a very long time—and a lot of us, me a part of it—about ripping down industrial civilization and starting again. But the trouble is, that when industrial civilization falls over, we lose the global dimming effect, which is all the pollution in the sky. [They will] fall out of the sky and more solar radiation will get through to the planet and put our extinction level event on steroids. There's different statistics where people aren't sure how much warming is in the global dimming, but it's somewhere between a degree and a half and three degrees of warming. Which is pretty much instantaneous if industrial civilization fell over tomorrow. Most of those pollutants would be out of the atmosphere in week. So we could have three degrees of warming crash across the planet in a week or two. and that would wipe out global grain production on all the continents.

So, you know, things are going to change very quickly and people are underestimating the speed that this thing can happen at. A lot of people will say, "Oh, that's Hester being too dire and doomy." But excuse me, I've done sixteen ocean passages on small yachts and one of the things that a skipper on a small yacht does is follow the precautionary principle. When you see a storm a building, you prepare for the worst possible storm. And if it doesn't come as bad as you prepared for, you are sweet. But if you're not prepared, you'll just get knocked down, knocked over, and knocked out. Capitalism is in complete denial about abrupt Climate Change. Even the people running he big NGOs and the big green movement are still in denial about abrupt Climate Change. So we're just about to get hit by the perfect storm and no one's prepared for it. It's going to be ugly beyond comprehension.

SONNENBLUME: It's hard to even imagine.

HESTER: It is hard to imagine and a lot of people prefer not to go there. They'd prefer not to dwell on it and to hope for the best. But hope isn't a strategy.

SONNENBLUME: No, hope is not a strategy at all.

HESTER: Hope is the prerogative of the privileged. If you talk to mothers in Yemen or in Gaza who have the military industrial complex raining down on them as you and me are having this conversation, they're not living in the world of hope. They move their children into the basement and when there's a quiet period they try to get the kids to get some sleep so they can have some sanity. But they're not giving those children false promises. They're not saying we're going to be okay. They're saying we hope we'll be okay but we've got to prepare for the worst.

SONNENBLUME: There's so little knowledge in the United States and those things are even going on. I live on the west coast of the United States. I've been here for the last 17 years and there's a whole cultural strain here that believes in "manifesting reality through good intentions." What's always offended me about that is that, then if you're not doing well or not having "abundance," as they say—by which they mean material wealth—it's your fault because you don't have the right intentions. And I always think of the children in Fallujah who were born with grotesque birth defects because of the white phosphorous and depleted uranium that was used against that city in the early 2000's during the US invasion.

Children born with extra limbs, with the brain on the outside of the skull, this sort of thing. And it's like, well now wait a minute, what did they do wrong? Were they supposed to have purer intentions so that wouldn't happen to them, or what?

HESTER: That is such a bullshit perspective to be taking. It's a little bit like believing there's some white guy floating around up in the clouds who's looking down on us and making all the decisions. Well, if people want to believe in that nonsense, they can, but the reality on the ground is what most people are suffering from and going through, so I want to live in the real world and deal with what I see as happening rather than hoping that there's going to be some nice beautiful existence in another sphere after we go off our mortal plane. I firmly believe that I was born in heaven. I don't understand why Christians want to go to heaven when I was born there. I want to look after the one that I was born into. Show some respect.

SONNENBLUME: Well this whole thing of taking the divine or the sacred out of the earth, out of life itself, out of the plants and the animals and the people, and putting it up in the sky, in the next world, that whole idea came from those monotheistic religions in the Middle East that came out of the agricultural cities. They came from that first, quote, "agricultural revolution." So it's all a piece of it. Because you go back to the indigenous worldview again and there are still people living on this planet who don't look at it that way, who *do* see—as what you just said—that we are already in paradise. You don't have to look someplace else for what is holy. It is here.

HESTER: Yeah, and I find that very liberating. To already be there is a really great start in life. All these people have aspirations to go to this place and it's almost like they can't see the forest for the trees.

January 23, 2018

174

Cross-Country Sprint, Part 3

NINE MILE PRAIRIE

September 10th

I was born and raised in Nebraska, but only years after moving away was my curiosity piqued about prairies. As an ecosystem type, prairies exist in conditions too moist for desert flora and too dry to support forest. Grasses are the most prevalent family of plants, both in number of species and in sheer mass. Forbs—which are non-grass plants without woody stems (so, not trees or shrubs)—are less common but totally essential. Trees are rare except near water.

The prairies of the Great Plains formed two to five thousand years after the last glaciers retreated. Retreating ice left behind mixed sediments that were gradually built into topsoil over many centuries with the addition of wind-borne dust and decayed organic matter. The ecosystem co-evolved with various animals including Buffalo, Elk, Deer, Rabbits, and Prairie Dogs, the last of whom played an important role in aerating the soil and creating channels for water penetration with their extensive tunneling.

At one time, prairies dominated Illinois, southern and western Minnesota, Iowa, northern Missouri, southwestern Manitoba, both Dakotas, Nebraska, Kansas, Oklahoma, central Texas, southern Saskatchewan, southeastern Alberta, Montana, and the eastern parts of Wyoming, Colorado and New Mexico. At their peak, the North American prairies covered 677,394 square miles, an area nearly a quarter (23%) the size of the lower 48 states. Within this vast zone, Tallgrass prairies grew in the east, where it was wetter at a lower elevation, and Shortgrass prairie in the west, where it was drier and higher, with Mixed Grass prairie taking up a wide band in between.

Grasses might strike most people as boring, but the many species found in Tallgrass Prairies are not the stuff of lawns. Growing three to six feet high, they send down roots five to twelve feet. Though their vegetative portion is the most obvious part to us—and is indeed tall—the majority of the plant's bulk is underground; in the case of Big Bluestem grass, the volume of the roots is two to four times greater than the foliage. The masses of roots often form thick, perennial rhizomes that both spread horizontally and dig down deeply. How big can they get? According to Paul A. Johnsgard of the School of Biological Science at the University of Nebraska-Lincoln, the famous ecologist John Weaver "once calculated that a square foot of big Bluestem sod might contain about 55 linear feet, and an acre about 400 miles of densely matted rhizomes, from the surface to a depth of only a few inches." Johnsgard goes on to say: "The strong roots of big bluestem have individual tensile strengths of 55-64 pounds, making prairie sod one of the strongest of natural organic substances. It is indeed strong enough to construct sod-built houses that have sometimes lasted a century or more in the face of Nebraska's relatively inhospitable climate."[1]

All of this plant material, both above and below ground, ends up producing a lot of decomposed organic matter every season: about 3000 pounds per acre on the surface and 2000 per acre underground. The turnover rate—from production to decomposition—is also fast, taking about 15 months above ground and 3-4 years below. This contrasts greatly with ecosystems that feature more trees and shrubs; with such woody plants, turnover rate can be measured in decades.[2]

Fire played a crucial role in maintaining vegetation on the prairie by suppressing trees, returning nutrients to the soil, and clearing away vegetative detritus. Animals loved the fresh green shoots that popped up afterwards. Herds of Buffalo would travel hundreds of miles to graze such spots.[3]

Native Americans called prairie fires the "Red Buffalo" and they set them intentionally as part of their gathering and hunting activities. Given that the prairies are only five to eight thousand years old, and that humans have been living in the Americas for much longer, the role that the Native Americans played was possibly foundational. Some have described their actions as comprising "land management," but that term is problematic given the contrast between Native American and European relationships to the land, the former being more participatory and the latter more dominating, if not malicious. "Wildtending" is a better word.

The "opening of the West" to colonial settlement had a devastating effect on the prairie ecosystems and their denizens. The invaders who rushed into these lands in the 1800s, especially after the building of the cross-country railroad, plowed under the grasslands and hunted the Buffalo nearly to extinction.

The slaughter of the Buffalo herds by Europeans is an event of such enormous scale that I would characterize it as unimaginable. In 1800, these animals numbered in the range of 30-60 million. Accounts from that time describe herds of animals stretching to the horizon. Try to picture that. I can't. My eye knows only the colonized landscape: tilled under, chopped down, or raked over.

The commercial hide industry is what lead to the Buffalo's near extinction. Highly organized hunting parties killed hundreds, if not thousands, of animals every day. Hides were pulled off the carcasses by pounding a spike through the dead animal's nose and hooking it up to a team of horses. The remainder of the animal was left to rot. Later, impoverished settlers collected the bones which were shipped to factories for making fertilizer.

Profit was not the only motive. It was well known that the Native American tribes of the Great Plains depended on the Buffalo for food and that by wiping out the animals, you would be threatening the humans. Decisions at the federal level in Washington, D.C., supported this policy. When a government intentionally sets out to destroy a group of people based on ethnicity, that's the definition of "genocide."

By the 1880's the Buffalo had been reduced to a few hundred. Some were protected on private ranches by individuals wishing to save them. The last remnant of truly wild Buffalo hid out in a valley in Yellowstone National Park and numbered just 23 at its lowest point.[4]

From tens of millions to mere hundreds in a few decades: what kind of travesty is that? What kind of sickness has overtaken a people when they engage in that kind of behavior? This greed cannot be excused as human nature, since other people cohabitated with the creatures for millennia without acting the same way, and in fact expanded their range with their wildtending practices.

No, there is something special about Western Civilization, and I mean that in the worst way. The outbreak of brutality can be traced to the agricultural revolution, a dramatic shift that led directly to cultures based on hierarchical domination and to lifestyles dependent on widespread environmental degradation.

This new worldview took expression in the Biblical injunction in Genesis 1:28, to "subdue" the earth and to "have dominion over the fish of the sea, and over the fowl of the air, and over every living thing that moveth." Other translations replace "have dominion over" with "rule," "reign over," "be masters over" or "in charge of" and these are all synonyms so the top-down character of this relationship is not in dispute. Contemporary adherents to the Abrahamic religions who wish to recast their faith as environmentally responsible must reckon with this concept, which is at the heart of their traditions, and without which the moral of the story is very different.

Believe it or not, the Yellowstone Buffalo herd is still threatened. Though it grew throughout the 20th Century, and in the past decade its numbers have floated around four to five thousand, its individuals are not protected from being slaughtered. If they go outside the Park boundaries, which they tend to do every winter in search of food and calving grounds, they are at the mercy of the "Interagency Bison Management Plan," under which the Montana Department of Livestock and National Park Service harass, capture and kill Buffalo. Over 11,500 have been murdered under this program since 1985. The stated pretext is that Buffalo will endanger cattle by infecting them with brucellosis, but there has never been a single documented case of that happening. No matter what spurious excuse is put forward, the true motivation for the annual killing is much deeper and darker, and it is this: Western civilization is just *that* profoundly sick that it can't leave this last wild remnant in peace. It *must* torture it; it's in the cultural DNA. Witness this account, as posted by the Buffalo Field Campaign, an activist organization that defends the Yellowstone herd:

> *On March 7, 1997, during a winter when 1,084 buffalo were killed, American Indian tribal leaders from around the country gathered near Gardiner, Montana, to hold a day of prayer for the buffalo. The ceremony was disrupted by the echo of gunshots. Lakota elder Rosalie Little Thunder left the prayer circle to investigate the shots. Less than two miles away, Department of Livestock agents had killed fourteen buffalo. Walking across a field to pray over the bodies, she was arrested and charged with criminal trespass. To Little Thunder and other tribal members present there was no question of coincidence: "They shot the buffalo because we were at that place on that day at that time," she said.[5]*

As the Buffalo were being decimated, the entire floristic web of the Tallgrass Prairie was being plowed under. Unfortunately for the prairie community, the soils it produces are ideal for agriculture. Explains Johnsgard:

> *The soils of Tallgrass prairie are among the deepest and most productive for grain crops of any on earth. They represent the breakdown products of thousands of generations of annual productivity of grass and other herbaceous organic matter. Because of these organic materials and the clays usually present in prairie soils, such soils have excellent water-holding capabilities. In addition to the humus and related organic matter thus produced, many prairie legumes have nitrogen-fixing root bacteria that enrich and fertilize the soil to*

a depth of at least 15 feet. Earthworms and various vertebrate animals such as gophers make subterranean burrows that mix and aerate prairie soils, in the case of earthworms to a depth of 13 feet or more.[6]

So for all the species of plants and animals of the Tallgrass Prairie—who number in the thousands if you count the insects—the end was nigh as soon as the Europeans arrived en masse, which didn't happen until after 1850. Less than a century later, most of this unique ecology was gone. In the present day, the Tallgrass Prairie ecosystem is even more rare than old growth forest, with less than 4% of it remaining. Most of that is in the western part of its former range, in Kansas and Oklahoma. In the eastern parts, such as Illinois, less than 1% is left. Like the slaughter of the Buffalo, a loss on this scale is unimaginable.

Agriculture replaces the wild with the domestic and in the Tallgrass Prairie, it did so rapidly, with deranged ruthlessness.

It was with all of this in the back of my mind—the Buffalo, the Native Americans, the untilled grasslands—that I visited Nine Mile Prairie outside of Lincoln, Nebraska. I was astounded, and in some way it was *the* highlight of the entire cross-country trip.

The preserve is owned by the University of Nebraska-Lincoln and is located five miles west and four miles north of its downtown campus; hence the name. It is only 230 acres in size, which is barely more than a third of a square mile, but even so it is one of the biggest parcels of "virgin" Tallgrass Prairie in the whole state.

Nine Mile Preserve is not set up as a tourist destination. The parking area only accommodates a handful of vehicles, the entrance is not well marked, and the only signage is well inside. This is all for the best, as the preserve's primary function is for research, and the fewer disturbances the better.

The first flower that greeted me inside the fence was the poorly named Purple Poppy Mallow (*Callirhoe involucrata*). Yes, it was undoubtedly a mallow, and certainly poppy-shaped, but "purple" was not right at all; it was much closer to "magenta," "fuchsia" or possibly "cerise" (according to crowd-sourced suggestions I solicited on social media). Regardless of shade, the blossom itself was striking: five rounded petals formed a cup, white at the center, was a dense cluster of light yellow anthers shedding pollen enthusiastically. The foliage was also distinctive: lobed like the familiar Maple, but with deeper indentations and sharper points. The shape could have been the print of some strange web-footed creature. According to Jon Farrar—whose *Field Guide to Wildflowers of Nebraska and the Great Plains*[7] I drew on for many of the IDs and much of the ethnobotany that follows herein—the roots are both edible (raw or cooked) and medicinal. The Teton Dakota Native Americans "inhaled the smoke of burning, dried roots for head colds" and drank a tea of boiled roots "for assorted internal pains."

But what to call *C. involucrata* since "Purple Poppy Mallow" won't do? Other common names include Claret Cup (in reference to the flower's shape), Buffalo Poppy (a tribute to the prairie's former inhabitants), Low Poppy Mallow (which describes its growth habit) and Cowboy Rose (a salute to the region's conquerors). Personally I prefer one of the first two, and would reject the last one out of hand; my heroes were on the other team.

Purple flowers that were actually purple soon appeared as we went further into the refuge: New England Asters (*Aster novae-angliae*), with their classic, daisy-style flowers; two Gayfeathers: Rough (*Liatris aspera*)—aka Button Snakeroot and Rattlesnake-Master—and Dotted (*L. punctata*)—aka Blazing Star and Starwort—with their spike-like inflorescences; and Purple Coneflowers (*Echinacea angustifolia*), renowned for their medicinal properties, and unmistakably identifiable by their drooping pinkish-purple ray petals and bristly collection of red-tipped disc florets.

Blue blossoms were displayed by only one plant that I saw: Pitcher's Sage, whose scientific name, *Salvia azurea*, nails it; "azure," the "color of the clear sky."[8]

White was represented by the low-growing, dainty-flowered Heath Aster (*Aster ericoides*); the erect, brushy-blossomed Fragrant Cudweed, (*Gnaphalium obtusifolium*)—aka Sweet Balsam, Rabbit Tobacco or Poverty Weed (this last perhaps referring to its affinity for disturbed settings); and the cotton-topped White Snakeroot (*Ageratina altissima*)—aka Deerwort Boneset and White Sanicle—a plant that struck back against the agricultural colonization of its home. To wit (as related by Farrar):

> *The herbage contains the toxin trematol. Snakeroot flourished, and was apparently more frequently eaten by cattle, when woods were cleared by pioneers and more attractive forage was unavailable. When passed on to humans in cow's milk, trematol causes milk sickness, the disease from which Nancy Hanks, Abraham Lincoln's mother, died.*[9]

Yellow was by far the most widespread color. Nebraska's state flower, the Goldenrod (genus *Solidago*, many species), was in full summery bloom. I was quite taken by the Showy Partridge Pea (*Chamaecrista fasciculata*), whose brightly colored flowers climbed its spindly branches between splays of feathery compound leaves and slender green seed pods, still flat in their fresh immaturity. I was delighted to meet Curly Cup Gumweed (*Grindelia squarrosa*), whose genus I had come to appreciate in the West—from the Smith River Delta to the Olympic Peninsula—for its intense flavor and medicinal potency. Like its relatives out there, this one was also used by Native Americans to treat various conditions: "colic, kidney problems, bronchitis, skin rashes, smallpox, pneumonia, gonorrhea, tuberculosis, and saddle galls on horses." Further, "powdered flower heads were used as asthma cigarettes by early settlers."[10] Intriguing! Smoking has largely fallen out of favor as a delivery method for medicinal herbs (with the exception of *Cannabis*) but I presume the mode has declined due to an unfair association with commercial tobacco products rather than a lack of efficacy.

The yellow flowers that attracted me most were the wild sunflowers. My surname, *Sonnenblume* is the German word for "sunflower" so I consider them siblings. There were many kinds there, but lacking a field guide, I couldn't identify them. I knew the genus—*Helianthus*—but what species was I seeing? *Grosseserratus* (Sawtooth), *petiolaris* (Plains or Prairie), *maximiliani* (Maximilian's), *tuberosus* (Jerusalem Artichoke, Canada Potato or Sunchoke), or the plain old garden variety *annuus*? Were there also *False* Sunflowers (*Heliopsis helianthoide*)? Maybe!

All these flowering plants were mixed in with grasses. Some held their heads above, some filled the spaces between the clumps, and still others found a niche close to the ground. The tallest grasses were over five feet high, and it was a joy to stand in the middle of a patch like that. I tried to imagine what it was like when this was all you could see.

But when I raised my eyes from my immediate surroundings, I saw the fencing around the preserve and beyond that fields, roads, and buildings. In the distance was the state capitol building in downtown Lincoln (about a nine mile drive away).

Unlike most state capitols, which are modeled on the US Capitol in Washington, DC, Nebraska's is centered around a tower. It was designed in 1920 and the architectural style is deemed to be "classical" but I would describe it as "proto-Deco" due to its sleek lines. Wikipedia notes that the 400 foot structure is sometimes referred to as, the "Tower on the Plains," and I'm sure that's true, but the nickname that I heard was "Penis on the Plains." Not for nothing does that moniker fit. The undeniably phallic tower is capped with a dome that is in turn topped with a statue of—I'm not making this up—a man sowing seed out of a bag. Surely if archaeologists dug up such a structure they would describe it as a temple to fertility worship, wouldn't they? And the difference here is... what, exactly?

Be that as it may, active processes of fertility and reproduction were going on at Nine Mile Prairie that day. Flowers contain the sex organs of plants and they are delivering an explicit "come hither" message to pollinators with their colors and shapes. That day I saw hundreds of insects: bees, butterflies, wasps, beetles and more. Specifically, I identified three butterflies: the Red Admiral (*Vanessa atalanta*), Sulfur (family Pieridae, subfamily Coliadinae) and Pecks Skipper (*Polites peckius*); a Soldier Beetle (*Chauliognathus pensylvanicus*); a Broad-headed Bug (genus *Alydus*); a male Golden or Northern Paper Wasp (*Polistes fuscatus*); and a Crab Spider (family Thomisidae). This is in addition to a multitude of pollinators including Bumble Bees. Where insects fly, web-spinning arachnids set up shop, and I found a big fat garden spider, striped yellow and black, stationed brazenly at the center of her net, which she had marked, running-light style, with zig-zags in stitched bold face.

Some plants were done flowering for the season, but I recognized their fruited or seeded forms: Common Milkweed (*Asclepias syriaca*) for its spindle-shaped pods; Wild Rose (genus *Rosa*) for its red hips; Ground Cherry, aka Tomatillo (*Physalis heterophylla*) for its husk-covered fruits; and Illinois Bundleflower (*Desmanthus illinoensis*), for its round, prickly seed pods. This last plant I had grown back in my farming days, hoping to process it for its dimethyltryptamine content, though I never got around to that.

Another plant I knew from farming whom I had never met in the wild —and which was the most exciting introduction of the day—was Compass Plant (*Silphium laciniatum*). I originally got seed from Richo Cech's Horizon Herbs (now Strictly Medicinal Seeds), of Williams, Oregon, because it was indicated for lung ailments. Richo has a way with words and describes the plant's appearance so: "Towering herbaceous perennial with deep delving roots... The large, handsomely and characteristically lobed leaves are very impressive, designed by nature to rise up through and push away prairie grasses. The stems are heavy, thick, hairy and green, glistening with fragrant and bitter gum." He later adds this advice: "During dormancy, burn off over the crown every few years (they won't mind, they are stimulated and cleaned by the fire and nourished by the ash)."[11] After years of ordering seeds and plants from his company, I came to know I could count on Richo's dry wit as well as his botanical expertise.

"Characteristically lobed" as a description of the leaves does understate the case, though. Here, Farrar goes the further mile:

Leaves leathery, clustered at base, up to 15 inches long and half as wide, deeply notched nearly to midrib forming lobes that are likewise notched. Stem leaves alternate, becoming progressively smaller, bases clasping stem. Leaves hairy but not conspicuously so, princi-pally along main leaf veins, rough to the touch.[12]

It was from this foliage that I recognized Compass Plant (who had bloomed earlier in the year). I snapped some photos with my phone and texted them to Clarabelle, knowing she would be excited too. We had grown the plant in Oregon and I always wondered if he (I thought) felt out of place and alone there. This was a plant integrated deeply into a particular ecological community; with the other plants and the animals to be sure, but also with the humid summers, the frigid winters and the wide open spaces. Over the years, when I reflected on the nearly complete destruction of the Tallgrass Prairies, I often mourned for the Compass Plant. I lovingly tended him in gardens, but when he shot up his towering flower stalk, I could see that he was missing his grass neighbors. I suppose all plants are shaped by their community, but *Silphium laciniatum* seems especially so to me. All the sadder that his home range is reduced to such ragged fragments.

Early conservationist Aldo Leopold, who witnessed the destruction of the Tallgrass Prairies first-hand in the early 20th century famously wrote, in his book, *A Sand County Almanac*: "What a thousand acres of Silphiums looked like when they tickled the bellies of the buffalo is a question never again to be answered, and perhaps not even asked."

I'm asking.

THE SANDHILLS

September 14th

Most people who have driven through Nebraska remember it as flat and boring farmland and if you take the east-west I-80, which most people do, then yeah, that's mostly how it is.

There's more to see than that, but you've got to follow the smaller roads. I knew this from family vacations in my childhood, when we explored much of the northern and western parts of the state. One of my most vivid memories of these travels was visiting the Sandhills. I can clearly recall treeless rolling hills covered with grass, ponds in the crevices between the hills, exposed sand on the steeper slopes. The place has always stood out in my mind as one of great beauty and uniqueness. The beauty is a matter of taste that others might not share, but the uniqueness is real.

The Sandhills region is the largest formation of sand dunes in the entire Western Hemisphere. At roughly 265 miles long, east to west, and 125 miles wide, north to south, it is about 20,000 square miles in size (estimates vary depending on source). Individual dunes can be as tall as 400 feet and as long as 20 miles. Most people think of dunes as mounds of exposed sand constantly shifting in the wind and while this was the state of the Sandhills at one time in their history, they are now 95% covered with vegetation, most of which is grasses. The region belongs to the Tallgrass or Mixed Grass prairie depending on who you ask, but is best understood on its own terms.

18,000 years ago, during the last Ice Age glaciation, the area was home to boreal Spruce forest. That's a signifier of how different the climate was then; in our current time, boreal forests—aka "snow forests" or "taiga"—are found in Canada. Further back still, the area was an inland sea. On one of those family vacations, we visited the Agate Fossil Beds, only fifty-some miles west of the Sandhills, where a trove of marine creature fossils can be found.

When the glaciers receded, they left behind quartz sand in the region. During an extended drought about 8000 years ago, dunes were formed from this material under the influence of prevailing north winds, making east-west lines that are still visible today in satellite imagery. Wetter conditions followed and the dunes were stabilized by vegetation. Two subsequent waves of drought and dune formation further shaped the topography, though each phase was weaker than the previous. Most of the area is currently stabilized, with the exception of the western edge. If global warming leads to pronounced drought and an attendant loss of vegetation, then the dunes could once more be set in motion.[13]

The Sandhills are not "pristine." However, they are the least disturbed area in all the Great Plains. The sandy soils are too unstable for farming (which people found out the hard way) so the vast majority remains unplowed. Ranching is widespread, though, and undeniably impacts vegetation (from being overgrazed), wildlife (from being hunted and displaced) and waterways (from being dredged, dammed and drawn down). Additionally, wildfires have been suppressed, which excludes a key element of cyclical regeneration.

At least 670 species of native plants have been found in the Sandhills, plus 50 non-natives ones. However, given the young age of the ecosystem, the majority of the plants there are found elsewhere. One exception is the Blowout Penstemon (*Penstemon haydenii*), native only to the Sandhills and limited parts of Wyoming. Though once abundant, the plant is now quite rare because of habitat loss to ranching.[14]

I spent one long day driving through the region east to west on Highway 2, which shadows the railroad much of the way. All day long, coal trains were traveling west to east on the tracks, one after the other. Not literally bumper-to-bumper—or caboose-to-engine—but quite often. It was a lot of coal. I felt a little sick inside, imagining the landscapes destroyed by mining and the pollution from its burning. I thought, too, about how the raping and pillaging of the environment remains unseen to most people, because so much of it is happening in "flyover land."

If the resource extractors have their way, coal won't be the only fossil fuel being shipped through the heart of the Sandhills. The Keystone XL pipeline is also supposed to run across the region. Local opposition has included people from all over the political spectrum, in part because of the violation of property rights by the use of eminent domain along the proposed route. Still, once different people get together over common interests, sometimes they find their commonalities and their interests expanding. In at least one well-publicized account, a white farmer deeded his land back to the Indians as part of the effort to halt the pipeline.[15]

Hidden from sight beneath the Sandhills is something else that's threatened by the pipeline and by other destructive practices of industrialized society, and that's the Ogallala Aquifer, an enormous natural reservoir of underground water. It's a relic of the last ice age, holding the melt-water of the glaciers, and kept topped off with rainwater since then. Though average precipitation in the Sandhills is low—23 inches per year at the eastern end and 17 inches at the western—the high porosity of the soils leads to nearly all of it being absorbed and added to the aquifer rather then running off.

The resulting high water table is responsible for the thousands of ponds, lakes and wetlands that dot the Sandhills, especially in its northwestern section. These water sources, though some are seasonal, attract multitudes of migrating birds and provide principal stops on the Great Plains central flyway. According to counts from the National Wildlife Refuges in the region, at least 270 species of bird pass through the area or reside there permanently. This includes the endangered Whooping Crane, whose numbers fell to less than two dozen early in the 20th Century.

However, since the 1950's, the aquifer has been suffering depletion chiefly due to the wide-scale implementation of center-pivot irrigation for agriculture. The huge circles that mark the landscape of the western states visible from planes or in satellite images are made by these giant systems. Often they are running in the middle of the hot day although that results in a high loss to evaporation. But most farmers consider themselves conservative and perversely that means they are not interested in conservation. Make sense of that one.

Nearly all the land in the Sandhills is privately owned. This is an outcome of the Kincaid Act of 1904 when homesteaders could claim land in

640 acre sections. Close to 9,000,000 acres were scooped up between 1910 and 1917 this way.[16] When I wanted to stop and experience the landscape, wide spots in the road were the primary places to do so. There are a number of wildlife preserves that are open to the public but they were not located along my route. Highway 2 has very little traffic, though, so these stops were peaceful enough. If I wanted to take a photo while standing in the middle of the road, that was no problem.

A place that was magical as a child can be mundane when revisited as an adult, but not so with the Sandhills for me. The sense of awe on this trip was just as great or more. I was mesmerized by the lines of the soft-edged hills undulating, ridge after ridge, into the distance, and the mystery of what lay between them, hidden from my eye. The play of shadow and light on the grassy slopes as clouds skimmed across the immense sky affected thrilling contrasts. The sheer expansiveness of the broad vistas stretched my senses. Close at hand, I appreciated the details: flowers, insects, smells and sounds, all of them wild and enlivening; barely met, but loved no less.

At one stop, I spotted a caterpillar crossing the road. I carefully picked him up and carried him to safety on the other side.

A moment later, a truck drove by and I was thankful I had acted.

NEVADA

Sept. 26th-27th

I crossed the state-line into Nevada a few minutes after sunset. The sky was glowing pinkish-orange beneath a deepening blue and the mountains ringing the horizon were fading to dark smudges. My first notice that I'd left Utah was a red, white and blue sign for the "Border Inn," which offered a motel, RV Park, Cafe, Food store, and—in the largest font—SLOTS, "open 24 hours." I was confused for a moment. I knew I'd been getting close to the border, but I hadn't seen a "Welcome to Nevada" sign, and this was definitely not a Utah establishment. Then I spotted it, over on the other side of the road, but it was unlit and much less prominent than the Border Inn's beacon, which had so ostentatiously alerted me that I was no longer under Mormon thumb and could start gambling right this minute. I chuckled to myself.

I stopped to check the map. With the light fading fast, it was time to find a camping spot, especially since I was on a two-lane road. I can be easily blinded by oncoming headlights and try to avoid driving under those conditions. Fortunately for me, just a few miles up the road was the

Sacramento Pass Recreation Area, a campground managed by the BLM (Bureau of Land Management).

By the time I arrived, darkness was setting in. I quickly parked and had just set up for the evening when the full moon appeared, very bright. She was rising through horizontal bands of cloud shaped like torn strips of semi-translucent paper. As she passed behind different layers, her disc was variously obscured, clarified, diffused, or intensified, before finally breaking out into the clear sky, where her glow dimmed the nearby stars above and cast sharp shadows on the ground.

Most people live in cities these days, so they don't know that moonlight casts shadows. Urban light pollution overwhelms her radiance, bathing every building, street and person in an artificial cast that banishes not just darkness but luminosity too. Most people don't know that a half moon provides enough light to find your way and that a full can enough to read by. I myself had no idea until I moved out of the city and started observing these things more closely.

In the morning I hit the road, embarking on that section of Highway 50 known as "The Loneliest Road in America." It takes you through only four towns—Ely, Eureka, Austin, Fallon—between the Border Inn in the east and the junction with Highway Alt-50 to Reno in the west—a distance of some 330-odd miles. But it's "lonely" only if you don't find good company in the basins and ranges that compose the landscape, which *I* do, personally.

The majority of the state of Nevada lies in a geographical area known as the Great Basin, which also takes up nearly half of Utah and a fair chunk of Oregon. Its western border is in California, at the base of the Sierra Nevada mountains, and it nibbles at a corner of Wyoming on its eastern edge. "Great Basin" denotes the fact that none of its watercourses within it run to the sea.

Highway 50 alternates between straight shots across flats and sections of winding switchbacks over the mountains. At the peak of each pass, another basin comes into view, bounded by another ridge beyond it, and down you go, with many twists and turns, to find the next straight shot. The shapes of these places strongly evoke the effects of running water and sure enough, the basins held lakes as recently as 12,500 years ago. According to Donald Grayson, author of *The Great Basin: A Natural Prehistory*, the region had at least 27,800,000 acres of lakes during the late Pleistocene (125,000 to 11,500 years ago), as compared to about 2,500,000 today.[17] That's well over ten times as much water. That landscape would have been laced with rivers, streams and springs and supported wetlands.

We often forget that the landscapes we see are not as they have always been. We are surprised when a river bend shifts course, a landslide buries a road, or a cliff-side crumbles into the sea, but these are minor incidents, akin to you or I trimming our nails. Geologically, on a planet whose history spans at least four billions years, 12,500 years is the blink of an eye, and the drift of continents, raising of mountains, and draining of inland seas are unexceptional events. Were we more attuned to this reality, perhaps we'd be more humble. It's not too late to try to cultivate such perspective. I suspect we would be reviving a consciousness we formerly enjoyed.

Humans have been living in the Great Basin for at least 12,000 years. Agriculture was never practiced there on a wide scale, like it was east of the Mississippi or in Central America, Even when some horticultural crops were cultivated, the gathering of undomesticated plants and hunting of wild animals remained central to the lifestyle. The most recent Native American residents to migrate into the area were the Shoshone, who still live there, though their territory has been shrunk and they are forbidden their traditional gathering and hunting activities at the levels they previously practiced them.

One of the staple foods of Native Americans in the Great Basin for millennia has been the pinenut. All pine trees produce pinenuts in their cones, but some are definitely better than others for eating due to their size, flavor, or ease of shelling. One of the best in the world is the Single-Leaf Pinyon, known scientifically as *Pinus monophylla*. Highway 50 passes through the heart of Single-Leaf Pinyon country,. Mixed in with Juniper trees, these forests grow on the slopes of the mountains starting at a particular elevation, so they are distinctly visible as dark, horizontal bands across the ranges.

US Americans did not settle in Nevada in large numbers until the 1870's. Precious metals were discovered there and thousands of square miles of Pinyon-Juniper woodland was clear-cut and the trees burned in giant kilns to make charcoal for processing gold and silver ore. As with the slaughter of the Buffalo on the Great Plains, the eradication of Pinyon trees in the Great Basin denied Native Americans a mainstay of their diet.

After World War II, a second deforestation effort was implemented to convert forest to pasture for cattle. In the current day, Pinyon and Juniper trees who are reclaiming their original ranges from these incursions have been declared, "invasive" and are being destroyed in a third malicious wave. The real motive for their removal is to benefit the ranching industry, which is comprised of invasive humans raising invasive cattle.[18]

Single-Leaf Pinyons don't produce the same amount every year, and can skip whole seasons, so if you want to harvest any, it's best to check out the groves ahead of time to see where the best spots will be. In the past, I had gathered at a particular place off of Highway 50, so I stopped there to see how it was doing. The location is on National Forest land near the town of Austin, and according to Forest Service rules, it is legal to collect a certain amount of pinenuts for personal use without a permit (which commercial operations are required to have). I don't have much experience with the tree, and I could not tell if the cones I was seeing would be decent producers that year or not, so I took photos to share with friends who would want to know.

Three years previously, I participated in a pinenut harvesting camp at this spot. One afternoon, I spent some time gathering by myself. I was squatting on the ground, picking up pinenuts, and reflecting on the thousands of other people who had done the same, for thousands of years, in that same woodland. A vision flashed before my eyes: I saw the mark of every human footfall that had ever touched the ground around me. They were countless in number, overlapping, pointing every direction, covering every square inch. In that instant, I understood that I was not alone; by engaging in this particular activity at this particular place, I was tied to all the other people who had done it there before me (and, one could surmise, to everyone who would).

The vision came and went in a blink. But it was crystal clear and I can still see it in my mind's eye. My sense was that the place itself held these marks of memory. Imagine tuning your senses to pick up on these impressions everywhere. They might also manifest as sound, smell, flavor or touch.

What does this tell us about place, that it can hold and transmit such energy? Does this not mean that places are "alive"? Some might roll their eyes at the notion, and my response is that of course a place is not alive in the same way that a contemporary human is. First off, a place is not limited to such a short-lived mortal form, and secondly, a place is not so ruled over by ego. So no, naturally not.

We must remember that the scope of our encounter with life is limited by our rudimentary senses and our narrow perspectives. But only the first is endemic to our existence (and as such cannot be deemed defective). The second, on the other hand, is cultural, and so can be transcended.

The Great Basin did not always look like this. Once, these salty plains were lakes and these mountains were islands. We humans have not always lived the way we do now. Once, we walked under the trees with senses open and outlooks wide. In the long cycles of geology, water will fill the basins again before too long and in the brief moment that is one human life, we all have the ability to attune again to the eternal. Perhaps, like the cycle of seasons, that too is inevitable, if only with our last breath.

June 2019

NOTES:

[1] Johnsgard, Paul A. "A Guide to the Tallgrass Prairies of Eastern Nebraska and Adjacent State" (2007) *Papers in Ornithology*, 39, pp. 2-3.

[2] *ibid*, p. 3

[3] Tallgrass Prairie National Preserve, Kansas. "Fire Regime." https://www.nps.gov/tapr/learn/nature/fire-regime.htm [retrieved 7/8/19]

[4] Wikipedia. "Buffalo Hunting."

[5] Buffalo Field Campaign. "Yellowstone Buffalo Slaughter Then and Now." https://buffalofieldcampaign.org/yellowstone-buffalo-slaughter-history [retrieved 7/8/19]

[6] Johnsgard, p. 3.

[7] Farrar, Jon. *Field Guide to Wildflowers of Nebraska and the Great Plains*, Second Edition (Iowa City, Iowa: University of Iowa Press, 2011).

[8] Merriam-Webster.

[9] Farrar, pp. 88-89.

[10] Farrar, p. 117.

[11] Strictly Medicinal Seeds. "Compass Plant (Silphium lacinatum), packet of 20 seeds, organic." https://strictlymedicinalseeds.com/product/compass-plant-silphium-laciniatum-packet-of-20-seeds-organic/ [retrieved 7/8/19]

[12] Farrar, p. 134.

[13] Zupancic, John W. "Formation of the Nebraska Sand Hills" (Prepared in partial fulfillment of requirements for the course in Remote Sensing, ES 771 at Emporia State University, December 6, 2001). http://academic.emporia.edu/aberjame/student/zupancic2/Page1_home.htm [retrieved 7/8/19]

[14] UNL Institute of Agriculture & Natural Resources, West Central Research & Extension Service. "The Nebraska Sandhills." https://extension.unl.edu/statewide/westcentral/gudmundsen/sandhills/ [retrieved 7/8/19]

[15] Hefflinger, Mark. "In Historic First, Nebraska Farmer Returns Land to Ponca Tribe Along "Trail of Tears"" (Bold Nebraska, June 11, 2018). http://boldnebraska.org/in-historic-first-nebraska-farmer-returns-land-to-ponca-tribe-along-trail-of-tears/ [retrieved 7/8/19]

[16] Spirit of Nebraska Pathway. "Nebraska Sandhills." https://nebraskaeducationonlocation.org/natural-attractions/sandhills/ [retrieved 7/8/19]

[17] Grayson, Donald. *The Great Basin: A Natural Prehistory* (Berkeley: University of California Press, 2011) p. 94.

[18] Hill, Nicole Patrice & the author: "The Troubles of 'Invasive' Plants" (Macska Moksha Press, 2019). http://macskamoksha.com/2019/01/invasive-zine [retrieved 7/8/19]

Home as a Season, Not a Place

On May 24, 2019, I saw a Painted Lady butterfly (*Vanessa cardui*) near the town of Dinsmore, California. That's in the northeast corner of Humboldt County, near the Trinity County line, just off the 36, on the edge of Six Rivers National Forest. The terrain is hilly, fairly steep and treed with Firs and Oaks of various species. Though this spot was only a little over 50 miles from the 101 to the west, it takes close to an hour and a half to drive there, due to tight curves and steep climbs. It's out in the boonies for sure.

The butterfly had seen better days. One of her/his lower wings was over half gone and the others were a bit raggedy-edged. I had seen thousands of them every day for weeks earlier this year in southern California, especially in Anza-Borrego State Park. For awhile, I couldn't drive anywhere without my windshield getting streaked with yellow and my grill filling up with dead bodies. Try as I might, it just wasn't possible to avoid hitting them, there were so many. I felt a little less bad about it one day when I came out of a store and saw a couple of small birds on my bumper, gorging on the corpses. Other times, sitting in the desert, a cloud of them would pass through, literally hundreds per minute, alighting on the many, many flowers and passing on. (A "superbloom" was in effect too.) For as many as there were, it was tricky to get good photos. They were shy of close human presence and never landed in one place for long. What I had to do was a pick a good spot and sit still with my camera turned on and pointed, waiting for one to flutter within focusing distance.

Now, months later, I wondered idly if I had actually seen this very individual butterfly before, hundreds of miles away.

"Hello friend," I said. "Have we met?"

As I ruminated on the idea that this butterfly might have followed the weather from the southern desert to the northern forests, I suddenly received a picture. It came and went like a flash, but conveyed a tremendous amount.

One could term such a moment an "epiphany." Like others I've been blessed with, this one was not about facts, but comprehension. It was a personal "aha!" moment and as such I cannot share it with you entirely. Only an aspect of it translates easily into English words, but that goes something like this:

"Home" for some creatures is not a place but a set of conditions on the move; what could be called a "season." The season is like a wave rippling across a landscape. The scale can range in size from a single mountain slope to an entire continent. Though certain characteristics of a season can be measured—such as temperature and humidity—as a home it is a form of consciousness. One is carried by it and also carries it within. Far more than a mere worldview, it is in fact a world. The world for the many, many creatures who live there full-time.

Like a sound, a season has an attack and a decay. Some are soft, others harsh, and each has a different degree of reverberation or resonance.

As the wave of a season rolls through an area, multi-faceted relationships are sparked: an insect hatches from a chrysalis, a flower blossoms, a tree's tips green up. With other waves, a caterpillar spins a cocoon, a plant dies back to its root, and a crown of leaves turns golden.

A season has no hard-edged beginning or end. Nonetheless, it is clear and unmistakable.

In any given solar year, the exact timing of a season relative to the calendar will vary, but the season is the one with real significance. Just as the map is not the territory, the calendar is not the time. A gardener or farmer who plants by dates and not by signs isn't paying real attention.

The particular (and peculiar) location-centered view of most contemporary humans is totally at odds with seasonal living; in fact, the two are mutually exclusive. We spend so much effort acquiring a domicile for ourselves and such a substantial percentage of our time working to keep it and we are culturally required to do so because nearly everything is owned rather than common. We must buy our right to have someplace to breathe, eat and sleep. We are paying to live. It's obscene. And the system that upholds it is ecocidal.

The day after I saw the butterfly, I drove south into Mendocino County. I passed through several zones, starting from a spot that is definitely on the chilly side for the greater area—elevation and latitude hold it cooler and wetter—and ending in one that was noticeably warmer and drier.

Most of my journey followed the 101, which alternates between two-lane roads with curves posted at 30mph or less and four-lane freeways where the traffic roars through at 70+. The way is mostly through valleys, some narrow and overhung with trees, others wide and tilled for crops. Redwood remnants pop up occasionally, including the well-known (and once again—unbelievably!—threatened) Richardson Grove.

When I descended into the valley where Laytonville sits in a small sea of agricultural land, the California Poppies were popping. Though a few have been blossoming for the last month and a half here and there, this was now a full-on event—an obvious season. The density of flowers had undeniably increased from the last time I drove through the area, ten days previously.

My thoughts traveled to SoCal back in March and April, when the Poppies were peaking there, in the basins, on the slopes, and in the meadows. Now it was happening here. Their season had arrived. In this place, which was farmland rather than desert, the flowers were limited to the roadsides because the fields were either planted with crops or had been mowed short, which is too much disruption even for this genus of plant, who definitely has an affinity for disturbed sites. (A roadside is a textbook example of a disturbed site.)

California Poppies are in the genus *Eschscholzia*, which contains at least a score of species and subspecies. According to botanist Michael L. Charters on his website, calflora.net, this mouthful of a name is derived from Dr. Johann Friedrich Gustav von Eschscholtz (1793-1831), "a Latvian or Estonian surgeon, entomologist and botanist who came with the Russian expeditions to the Pacific coast in 1816 and 1824." I mention this because the Western habit of naming things—plants, animals, mountains, rivers—after people is a) a form of colonization, b) reflective of the out-of-balance ego of European culture, and c) not at all helpful in telling us anything about the thing named. Not only is the methodology not "scientific" but it's another way of severing us from living reality.

Unlike the Painted Ladies, who moved with the wave, the Poppies are stationary, reacting as the wave washes over them. How different is it, though? It struck me that all/every *Eschscholzia* might inhabit a collective consciousness that responds as the season passes through it, kind of like when a tingle passes through your body from head to toe. Perhaps this consciousness operates as a "we," but a we with no "I"s. That is, maybe California Poppy is awake and aware of itself as a single, vast, selfless self-ness whether blooming in a weed-filled ditch outside the town of Willits in Mendo in May or in a sandy wash at Mountain Palms Springs campground in Anza-Borrego in March. That makes the experienced "home" of Cal-

Poppy-in-flower a migratory event of sorts: a "season" independent of place, though individual plants are literally rooted in particular chunks of soil.

In the town of Willits, the explosion of blossoms was even more showy. There, they were weeded and watered, though they excel at taking their own space when its their moment anyway. Along the sidewalks and driveways, their bright orange and yellow faces were reminders of the world outside civilization, of the timing separate from clocks.

So caught up are most of us in our efforts just to survive here—to establish our place—that we forget that other ways of living exist. Yet once upon a time, before the Agricultural revolution, and for many, many millennia, we too lived and thrived in season. We were more like the Painted Lady, moving with changing conditions. We did not make as sharp of a distinction between ourselves and other living creatures in those days, nor between internal and external. We were far more collective than individual. We named things for their own characteristics.

Civilization has taken us very far from all of that, to the edge of a cliff where we are now surveying the possibility of our own extinction. Not merely our actions need to change, but our worldview; and not merely our worldview but our state of consciousness. That's way beyond the realm of politics and even philosophy. We must look within ourselves, deeply, and find again that spirit which rides the wind and opens to the sky in tune with the seasons.

May 26, 2019

About the Author

Kollibri terre Sonnenblume is a writer, photographer, tree-hugger, animal lover and political dissident who has been living on the West Coast of the United States since 2001. Kollibri farmed from 2005-2014, making a living with an urban CSA and a seed & herb company.

Previous books include "The Failures of Farming and the Necessity of Wildtending" (a collection of essays), "A Photographic Love Letter to the Flora and Fauna of the Mojave Desert" (a picture book), "Adventures in Urban Bike Farming" (a memoir), and "Wildflowers of Joshua Tree Country" (a botanical field guide). Earlier experiences include Indymedia activist, music critic and intellectual provocateur. Kollibri earned a Bachelor of Arts degree in "Writing Fiction and Non-fiction" from the St. Olaf Paracollege (Northfield, Minnesota) in 1991.

Keep up with Kollibri's latest at <u>macskamoksha.com</u>

Printed in Great Britain
by Amazon